# ENERGY SCAVENGING FOR WIRELESS SENSOR NETWORKS
## with  Special Focus on Vibrations

# ENERGY SCAVENGING FOR WIRELESS SENSOR NETWORKS
## with  Special Focus on Vibrations

*by*

**Shad Roundy**
**Paul Kenneth Wright**
**Jan M. Rabaey**
*UC Berkeley*

**KLUWER ACADEMIC PUBLISHERS**
**Boston / Dordrecht / New York / London**

Distributors for North, Central and South America:
Kluwer Academic Publishers
101 Philip Drive
Assinippi Park
Norwell, Massachusetts 02061 USA
Telephone (781) 871-6600
Fax (781) 871-6528
E-Mail <kluwer@wkap.com>

Distributors for all other countries:
Kluwer Academic Publishers Group
Post Office Box 322
3300 AH Dordrecht, THE NETHERLANDS
Telephone 31 78 6576 000
Fax 31 78 6576 474
E-Mail <orderdept@wkap.nl>

   Electronic Services <http://www.wkap.nl>

**Library of Congress Cataloging-in-Publication**

CIP info or:

Title: Energy Scavenging for Wireless Sensor Networks
Author (s):  Shad Roundy, Paul Kenneth Wright  & Jan M. Rabaey
ISBN: 1-4020-7663-0

# Dedication

*This book is dedicated to
Michelle, Caleb, and Seth*

# Contents

# Preface

The vast reduction in size and power consumption of CMOS circuitry has led to a large research effort based around the vision of wireless sensor networks. The proposed networks will be comprised of thousands of small wireless nodes that operate in a multi-hop fashion, replacing long transmission distances with many low power, low cost wireless devices. The result will be the creation of an intelligent environment responding to its inhabitants and ambient conditions.

Wireless devices currently being designed and built for use in such environments typically run on batteries. However, as the networks increase in number and the devices decrease in size, the replacement of depleted batteries will not be practical. The cost of replacing batteries in a few devices that make up a small network about once per year is modest. However, the cost of replacing thousands of devices in a single building annually, some of which are in areas difficult to access, is simply not practical. Another approach would be to use a battery that is large enough to last the entire lifetime of the wireless sensor device. However, a battery large enough to last the lifetime of the device would dominate the overall system size and cost, and thus is not very attractive. Alternative methods of powering the devices that will make up the wireless networks are desperately needed.

Alternative power methods could include improved energy reservoirs, such as improved primary batteries or fuel cells. They could include methods of wirelessly distributing power to the nodes, such as beaming RF energy to the target devices. Or finally, alternative power sources could include methods to scavenge power available in the environment of the node, thus making it completely self-sustaining for its entire lifetime. While the energy scavenging solution is perhaps the most difficult to implement, it also has the highest potential payoff because the lifetime of the node would only depend on the reliability of its parts.

This work reports on a research project to explore alternative power methods for wireless sensor nodes. While a large variety of power sources in each of the three categories just mentioned are evaluated, the focus is on the last category, energy scavenging methods. In the introductory chapter, the potential of a wide range of energy scavenging methods is explored. Recent research, if any exists, in each area is summarized. Based on this evaluation, low level vibrations have been identified as an attractive power source in many potential applications.

The remainder of the work is focused on researching vibrations as a power source. Commonly occurring vibrations are characterized and evaluated for the amount of power that could be scavenged from them. Different methods of converting the kinetic energy inherent in the vibrations to electricity have been evaluated and compared. Both piezoelectric and electrostatic MEMS generator devices have been modeled, designed, built, and tested. Power densities on the order of 200 $\mu$W/cm$^3$ have been demonstrated and wireless transceivers have been completely powered by low level ambient vibrations.

The hope is that the models and designs presented herein will help enable the application of vibration based generators in wireless sensor applications. Furthermore, this work can serve as a basis on which future research in the area can build.

# Foreword

Dear colleagues,

It is with great pleasure that I introduce this book by my associates Shad Roundy, Paul Wright and Jan Rabaey.

Our Center for Information Technology Research in the Interests of Society (CITRIS), was formed in the year 2000 to focus on a wide variety of challenging social issues that world faces today: better health care, increased food supply, clean water, improved global education, available energy supplies and protection from natural and man-made disasters. We are working as an integrated group of engineers, computer scientists, sociologists and lawyers to focus on these global challenges that urgently need solutions in this post-9/11 era.

We are deploying a wide variety of sensors, computers, information technology procedures, networking and wireless systems to address the above needs. Figure 1 shows the evolution of CITRIS's wireless sensor platforms. They are based on the "PicoRadio" project at the Berkeley Wireless Research Center (BWRC) and the "TinyOS" project at the Intel Berkeley Research Laboratory. The photographs show self-contained units that include the desired sensor(s), a small computer, a transmitter, and battery. Today's units, being used in preliminary experiments to monitor energy use in buildings, the deflections of the Golden Gate Bridge that spans San Francisco Bay, and the health of redwood trees in California forests are 2 to 6cm$^3$ in size (Figure 1b and 1c). In the future, this size (and of course cost) will reduce as research naturally evolves towards CMOS and MEMS-based systems. Figure 1d (just to the right of Lincoln's nose) shows the work of another CITRIS researcher, Kris Pister. Today we can fabricate fragile MEMS prototypes called "motes" because of their tiny size. Miniaturization will reduce costs further over the next few years with a potential size reduction to less than 1 mm³. As "smart dust" they may possibly and eventually be spray-painted onto the walls of air-conditioning ducts, or "injected" into the upholstery of furniture.

*Figure 1a to 1d: Evolution of <u>wireless, sensor "nodes" or" motes"</u> c: practical size of 2cm³ d: MEMS prototype*

As indicated above, the platforms in Figure 1a and 1b can carry a supplementary sensor board for a range of sensors. The meso-scale device (1c) has sensors permanently attached to the PCB. These include sensors for temperature, humidity, sunlight and artificial light, motion, and smoke. The platforms have also been connected to conventional strain gauges and accelerometers for measuring stress and vibration in Steve Glaser's earthquake studies.

But today there is a major snag to the easy deployment of such devices. Note the size of Figures 1a and 1b. Despite the small size of the wireless electronics, the final size is dominated by the battery packs. Worse yet, who is going to change all these batteries if the motes are deployed in a commercial building? No one probably! If you will forgive my informality, don't we all have a smoke alarm somewhere in our houses where we 'just don't get around to changing the battery'?

Therefore for platform power, CITRIS, led by the fine PhD work of Shad Roundy, has been creating Meso- and MEMS-scale **energy scavenging (or harvesting)** systems, driven by a hybrid of solar power, ambient vibrations, and other possible sources. For short platform-life, batteries are of course a reasonable solution, but the building industry, for example, prefers a >10year life. Solar cells offer excellent power density in direct sunlight, but in dim, or non-lit areas, they are inadequate.

*Figure 2. a: Meso-scale piezobender; b: Schematic and c/d: SEM of MEMS-energy-scavenging-from-electrostatics*

My colleagues have thus studied building vibration sources: for example, the internal metal ductwork is vibrating constantly. They have designed piezoelectric benders (shown in Figure 2a and throughout this book) to be mounted on vibrating surfaces, and shown that they generate 150-200 microwatts. For further miniaturization, we have preliminary designs for an electrostatic generator that can be created with MEMS processing (Figure 2b-d). While fundamental engineering-science is still needed on: 1) how to design power-train circuitry to collect energy at low rates while providing for high power "bursts" during network operation; and 2) how to store power in micro-capacitors or rechargeable on-PCB micro-batteries, and 3) how to optimize and prototype various vibration/solar hybrids, we are confident that this is an important technology for the robust deployment of sensor nets.

I hope you will enjoy reading this book and hence sharing with us the mission of CITRIS!

With best wishes

*Ruzena Bajcsy*

*Berkeley California 2003*

# Chapter 1

# INTRODUCTION
*Motivation and Potential Sources of Power*

The past several years have seen an increasing interest in the development of wireless sensor and actuator networks. Such networks could potentially be used for a wide variety of applications. A few possible applications include: monitoring temperature, light, and the location of persons in commercial buildings to control the environment in a more energy efficient manner, sensing harmful chemical agents in high traffic areas, monitoring fatigue crack formation on aircraft, monitoring acceleration and pressure in automobile tires, etc. Indeed, many experts foresee that very low power embedded electronic devices will become a ubiquitous part of our environment, performing functions in applications ranging from entertainment to factory automation (Rabaey *et al* 2000, Gates 2002, Wang *et al* 2002, Hitachi 2003).

Advances in IC (Integrated Circuit) manufacturing and low power circuit design and networking techniques (Chandrakasan *et al*, 1998, Davis *et al*, 2001) have reduced the total power requirements of a wireless sensor node to well below 1 milliwatt. Several low power wireless platforms with power consumption on the order of several to tens of milliwatts have recently become commercially available (Haartsen and Mattisson 2000, Dust 2003, Crossbow 2003, Xsilogy 2003, Ember 2003). Wireless sensor nodes developed in the research community are now boasting power consumption on the order of hundreds of microwatts (Hill and Culler 2002, Roundy *et al* 2003). Such nodes will form dense ad-hoc networks transmitting data from 1 to 10 meters. In fact, for communication distances over 10 meters, the energy to transmit data rapidly dominates the system (Rabaey *et al* 2002). Therefore, the proposed sensor networks will operate in a multi-hop fashion replacing large transmission distances with multiple low power, low cost nodes.

The problem of powering a large number of nodes in a dense network becomes critical when one considers the prohibitive cost of wiring power to them or replacing batteries. In order for the nodes to be conveniently placed and used they must be small, which places severe limits on their lifetime if powered by a battery meant to last the entire life of the device. State of the art, non-rechargeable lithium batteries can provide up to 800 WH/L (watt hours per liter) or 2880 J/cm$^3$. If an electronic device with a 1 cm$^3$ battery were to consume 100 μW of power on average (an aggressive yet obtainable goal), the device could last 8000 hours or 333 days, almost a year. Actually, this is a very optimistic estimate, as the entire capacity of a battery usually cannot be used due to voltage drop. It is worth mentioning that the sensors and electronics of a wireless sensor node will be far smaller than 1 cm$^3$, so, in this case, the battery would dominate the system volume. Clearly, a lifetime of 1 year is far from sufficient. The need to develop alternative methods of power for wireless sensor and actuator nodes is acute.

There are three possible ways to address the problem of powering the emerging wireless technologies:

1. Improve the energy density of storage systems.
2. Develop novel methods to distribute power to nodes.
3. Develop technologies that enable a node to generate or "scavenge" its own power.

This chapter will review the potential of many power sources falling into each of these three categories. The current state of research on each of the potential power sources will also be discussed. As the focus of this book is most particularly on vibration-to-electricity converters, particular attention is devoted to reviewing research done previously on vibration based generators.

A direct comparison of vastly different types of power source technologies is difficult. For example, comparing the efficiency of a solar cell to that of a battery is not very useful. However, in an effort to provide general understanding of a wide variety of power sources, the following metrics will be used for comparison: power density (μW/cm$^3$), energy density where applicable (J/cm$^3$), and power density per year of use (μW/cm$^3$/yr). Based on the current state of research in the field or wireless sensor networks, a practical level of acceptability would be an average power density of about 100 μW/cm$^3$ with a lifetime of 10 years. Additional considerations include the complexity of the power electronics needed and whether secondary energy storage is needed.

## 1. ENERGY STORAGE

Energy storage, in the form of electrochemical energy stored in a battery, is the predominant means of providing power to wireless devices today. In fact electrochemical batteries have been a dominant form of energy storage for the past 100 years. Batteries are probably the easiest power solution for wireless electronics because of their versatility. However, there are other forms of energy storage that may be useful for wireless sensor nodes. Regardless of the form of the energy storage, the lifetime of the node will be determined by the fixed amount of energy stored on the device. While it is cost effective in some applications to repeatedly change or recharge batteries, if wireless sensor nodes are to become a ubiquitous part of the environment, it will no longer be cost effective. The primary metric of interest for all forms of energy storage will be usable energy per unit volume ($J/cm^3$) and the closely related power per unit volume per unit time ($\mu W/cm^3$/year) of operation.

## 1.1 Batteries

Macro-scale primary batteries are commonly available in the marketplace. Table 1.1 shows the energy density and standard operating voltage for a few common primary battery chemistries. Figure 1.1 shows the average power available from these battery chemistries versus lifetime. Note that while zinc-air batteries have the highest energy density, their lifetime is very short. They are, therefore, primarily used in applications with fairly high and relatively constant power consumption, such as hearing aids. While lithium batteries have excellent energy density and longevity, they are also the most expensive of the chemistries shown. Because of their combination of energy density and low cost, alkaline batteries are widely used in consumer electronics.

Because batteries have a fairly stable voltage, electronic devices can often be run directly from the battery without any intervening power electronics. While this may not be the most robust method of powering the electronics, it is often used and is advantageous in that it avoids the extra power consumed by power electronics.

*Table –1.1.* Energy density and voltage of three primary battery chemistries.

| Chemistry | Zinc-air | Lithium | Alkaline |
|---|---|---|---|
| Energy Density ($J/cm^3$) | 3780 | 2880 | 1200 |
| Voltage (V) | 1.4 | 3.0 - 4.0 | 1.5 |

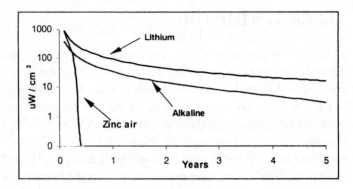

*Figure −1.1.* Continuous power per cm³ vs. lifetime for three primary battery chemistries.

Macro-scale secondary (rechargeable) batteries are commonly used in consumer electronic products such as cell phones, PDA's, and laptop computers. Table 2.2 gives the energy density and standard operating voltage of a few common rechargeable battery chemistries.

It should be remembered that rechargeable batteries are a secondary power source. Therefore, in the context of wireless sensor networks, another primary power source must be used to charge them. In some cases (such as cell phones or notebook computers) this is easily accomplished because the device can be periodically connected to the power grid. However, as wireless sensor devices become smaller, cheaper, and more widespread, periodically connecting the device to an energy rich source or the power grid will not be cost effective or may not even be possible. More likely, an energy scavenging source on the node itself, such as a solar cell, would be used to recharge the battery. One item to consider when using rechargeable batteries is that electronics to control the charging profile must often be used. These electronics add to the overall power dissipation of the device. However, like primary batteries, the output voltages are stable and power electronics between the battery and the load electronics can often be avoided.

*Table −2.2.* Energy density and voltage of three secondary battery chemistries.

| Chemistry | Lithium-Ion | NiMHd | NiCd |
|---|---|---|---|
| Energy density (J/cm³) | 1080 | 860 | 650 |
| Voltage (V) | 3.0 | 1.5 | 1.5 |

## 1.2    Micro-batteries

The wireless sensor nodes available today tend to be about the size of a small matchbox or pager. One such sensor node, the Crossbow Mica Mote

(Crossbow 2003), is shown in Figure 1.2. It is readily evident that the majority of the space is consumed by the batteries, packaging, and interconnects, not by the electronics. Therefore, in order to dramatically reduce the size of wireless sensor nodes, greater integration is necessary. A large research effort, with the ultimate goal of developing micro-batteries (or on chip batteries), is underway at many institutions. This section will review some of the issues in the development of micro-batteries and some of the research in the field.

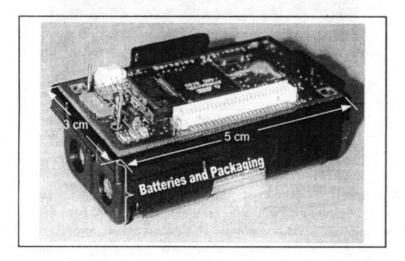

*Figure –1.2.* Mica Mote wireless sensor node from Crossbow (Crossbow 2003).

One of the main stumbling blocks to reducing the size of micro-batteries is power output due to surface area limitations of micro-scale devices. The maximum current output of a battery depends on the surface area of the electrodes. For low power sensor nodes using macro scale batteries, maximum power output is generally not an issue. However, because micro-batteries are so small, the electrodes have a small surface area, and their maximum current output is also very small. This problem can also be alleviated to a certain degree by placing a large capacitor in parallel with the battery capable of providing short bursts of current. However, the capacitor itself consumes additional volume, and therefore may not be desirable in many applications.

The challenge of maintaining (or increasing) performance while decreasing size is being addressed on multiple fronts. Bates *et al* at Oak Ridge National Laboratory have created a process by which a primary thin film lithium battery can be deposited onto a chip (Bates *et al* 2000). The thickness of the entire battery is on the order of 10's of μm, but the area is in the cm$^2$ range. This battery is in the form of a traditional Volta pile with alternating layers of Lithium Manganese Oxide (or Lithium Cobalt Oxide, LiCoO2), Lithium Phosphate Oxynitride and Lithium metal. Maximum potential is rated at 4.2 V with continuous maximum current output on the order of 1 mA/cm$^2$ and 5 mA/cm$^2$ for the LiCoO2 – Li based cell. A schematic of a battery fabricated with this process is shown in Figure 1.3.

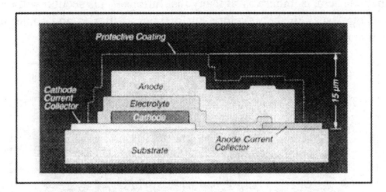

*Figure –1.3*. Primary Lithium on chip battery proposed by Bates *et al* (Bates *et al* 2003).

Work is being done on thick film batteries with a smaller surface area by Harb et al (Harb et al 2002), who have developed micro-batteries of Ni/Zn with an aqueous NaOH electrolyte. Thick films are on the order of 0.1 mm, but overall battery thickness is minimized by use of three-dimensional structures, which reduce the overall thickness because the cathode and anode are side by side rather than stacked one on top of another. While each cell is only rated at 1.5 V, geometries have been duty-cycle optimized to give acceptable power outputs at small overall theoretical volumes (4 mm by 1.5 mm by 0.2 mm) with good durability demonstrated by the electrochemical components of the battery. The main challenges lie in maintaining a microfabricated structure that can contain an aqueous electrolyte.

Radical three dimensional structures are also being investigated to maximize power output. Hart *et al* (Hart *et al* 2003) have theorized a three dimensional battery made of series alternating cathode and anode rods suspended in a solid electrolyte matrix. Theoretical power outputs for a three dimensional micro-battery of this type are shown to be many times larger than a two dimensional battery of equal size. Additionally, the three dimensional battery would have far lower ohmic ionic transport distances, and thus lower ohmic losses.

For example, a 1 cm$^2$ thin film with each electrode having a thickness of 22 µm and a 5 µm electrolyte, would have a maximum current density on the order of 5 mA. If the battery is restructured to have the same total volume, with square packing electrode rods (as Hart *et al* have proposed) of

5 μm radius with 5 μm surface to surface distance, geometry dictates that the energy capacity is reduced to 39% of the thin film capacity (due to a higher volume percentage of electrolyte for the thin film battery). However, while the energy density is lower for the 3D battery, the power density is higher due to a higher surface area. In fact, the three dimensional battery would have a total electrode area of 3.5 cm$^2$, an increase of 350%. The increase in surface area alone improves the current density to 17.5 mA. Moreover, the ionic transport scale in the 2D structure is about 350% longer than the 3D case because the electrodes for the 3D case are much thinner. Therefore, decreased ohmic losses could further improve the maximum throughput to 20 mA at 4.2 volts. However, the inherent non-uniformities in current distribution in three dimensional batteries (exacerbated by the particular complexity of this cell) may lead to difficulties with regard to device reliability on primary battery systems and cycle life in secondary battery systems.

## 1.3    Micro-fuel cells

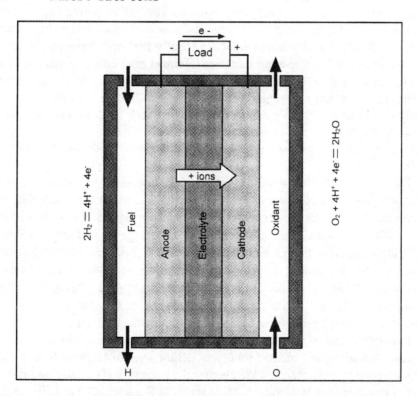

*Figure –1.4.* Illustration of how a standard hydrogen fuel cell works.

Hydrocarbon based fuels have very high energy densities compared to batteries. For example, methanol has an energy density of 17.6 kJ/cm$^3$, which is about 6 times that of a lithium battery. Therefore, fuel cells are potentially very attractive for wireless sensor nodes that require high power outputs for hours to days. Like batteries, fuel cells produce electrical power from a chemical reaction. A standard fuel cell uses hydrogen atoms as fuel. A catalyst promotes the separation of the electron in the hydrogen atom from the proton. The proton diffuses through an electrolyte (often a solid membrane) while the electron is available for use by an external circuit. The protons and electrons recombine with oxygen atoms on the other side (the oxidant side) of the electrolyte to produce water molecules. This process is illustrated in Figure 1.4. While pure hydrogen can be used as a fuel, other hydrocarbon fuels are often used. For example in Direct Methanol Fuel Cells (DFMC) the anode catalyst draws the hydrogen atoms out from the methanol.

Large scale fuel cells have been used as power supplies for decades. For example, the Apollo spacecraft used alkaline fuel cells for electricity. More recently, fuel cells have been developed as alternative power supplies for automobiles. Cells using a variety of fuels and electrolytes have been successfully used at the macro scale. Recently fuel cells have gained favor as a replacement for consumer batteries (Heinzel *et al* 2002). Small, but still macro-scale, fuel cells are likely to soon appear in the market as battery rechargers and battery replacements (Toshiba 2003). Holloday *et al* (Holloday *et al* 2002) have demonstrated a research fuel cell reactor with a total size on the order of several mm$^3$. The research trend is toward micro-fuel cells that could possibly be closely integrated with wireless sensor nodes.

Like micro-batteries, a primary metric of comparison in micro-fuel cells is power density in addition to energy density. As with micro-batteries, the maximum continuous current output is dependent on the electrode surface area. Microfabricated fuel cells offer an advantage in surface to volume ratio, thereby giving them a higher power density. Likewise microfabricated features can potentially improve gas diffusion and lower the internal resistance (Kang *et al* 2001), both of which improve efficiency. The millimeter scale fuel cell system by Holloday *et al* (Holloday *et al* 2002) produces roughly 25 mA from a thin cell with an area of 2 cm$^2$.

Fuel cells tend to operate better at higher temperatures, which are more difficult to maintain in micro-fuel cells. Efficiencies of large scale fuel cells have reached approximately 45% electrical conversion efficiency and nearly 90% if cogeneration is employed (Kordesh and Simader 2001). Efficiencies for micro-scale fuel cells will certainly be lower. The previously mentioned fuel cell by Holladay *et al* has demonstrated about 0.5% efficiency. They are targeting a 5% efficient cell. Given the energy density of fuels such as methanol, fuel cells need to reach efficiencies of at least 20% in order to be more attractive than primary batteries. Nevertheless, at the micro scale, where battery efficiencies are also lower, a lower efficiency fuel cell could still be attractive.

Most single fuel cells tend to output a voltage around 1.0 – 1.5 volts. Of course, like batteries, the cells can be placed in series for higher voltages. The voltage is quite stable over the operating lifetime of the cell, but it does fall off with increasing current draw. Figure 1.5 shows the voltage versus current load for a typical fuel cell. Notice that as the current density increases, the dominant loss mechanism also changes. Because the voltage drops with current, it is likely some power electronics will be necessary if replacing a battery with a fuel cell. Even if electronics are designed to run directly from a battery, they will probably not function well when powered directly from a fuel cell.

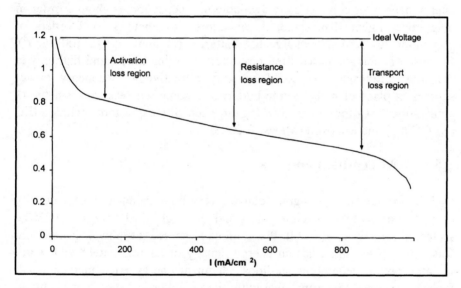

*Figure –1.5.* Typical voltage versus current curve for a fuel cell.

Finally, providing for sufficient fuel and oxidant flows is a very difficult task for micro-fuel cells. The ability to microfabricate electrodes and electrolytes does not guarantee the ability to realize a micro-fuel cell. The problem of microfabricating the fuel reservoir and all of the plumbing is arguably a more difficult task than the microfabrication of electrodes. To the authors' knowledge, a self-contained on-chip fuel cell has yet to be demonstrated.

## 1.4      Ultracapacitors

Ultracapacitors represent a compromise of sorts between rechargeable batteries and standard capacitors.    Standard capacitors can provide significantly higher power densities than batteries, however their energy density is lower by about 2 to 3 orders of magnitude. Ultracapacitors (also called supercapacitors or electrochemical capacitors) achieve significantly higher energy density than standard capacitors, but retain many of the favorable characteristics of capacitors, such as long life and short charging time.

Rather than just storing charge across a dielectric material, as capacitors do, ultracapacitors store ionic charge in an electric double layer to increase their effective capacitance.  By introducing an electrolyte researchers hope to limit ionic diffusion between plates, trading power generation for longer running times.  However, this is still an area of technical difficulty. The energy density of commercially available ultracapacitors is about 1 order of magnitude higher than standard capacitors and about 1 to 2 orders of magnitude lower than rechargeable batteries (or about 50 to 100 $J/cm^3$). Because of their increased lifetimes, short charging times, and high power densities, ultracapacitors could be very attractive as secondary power sources in place of rechargeable batteries in some wireless sensor network applications.  Corporations working on such ultracapacitors include NEC (NEC 2003) and Maxwell (Maxwell 2003).

## 1.5      Micro-heat engines

Like fuel cells, heat engines convert very high chemical energy density of hydrocarbon fuels to either electrical or mechanical energy.  At large scales, fossil fuels are still the dominant source of power generation. Gasoline, for example, has an energy density of approximately 31 $kJ/cm^3$. In large power generation facilities, chemical energy embedded in fossil fuels is converted to thermal energy through combustion, then to mechanical energy through a heat engine, and finally to electricity through a magnetic generator.  The complexity of this process, along with the many complicated parts has served as a barrier to the miniaturization of this technology. Recent progress made in silicon micromachining is removing this barrier by making it possible, and cost effective, to fabricate extremely small (on the order of microns) complex mechanical and fluidic structures.  As a result, there have been a number of recent research projects with the goal of a achieving a micro-scale heat engine.

To the authors' knowledge, the first such effort was proposed and undertaken by Epstein *et al* in the mid 1990's (Epstein *et al* 1997).  Epstein and colleagues designed and built high speed turbomachinery with bearings, a generator, and a combustor all within a cubic centimeter using a combination of silicon deep reactive ion etching, fusion wafer bonding, and thin film processes. An application ready power supply would additionally require auxiliary components, such as a fuel tank, engine and fuel controller, electrical power conditioning with short term storage, thermal management and packaging. The expected performance of the system proposed by Epstein *et al* was 10-20 Watts of output electrical power at thermal efficiencies of 5-20%. Figure 1.6 shows an example microturbine test device used for turbomachinery and air bearing development.

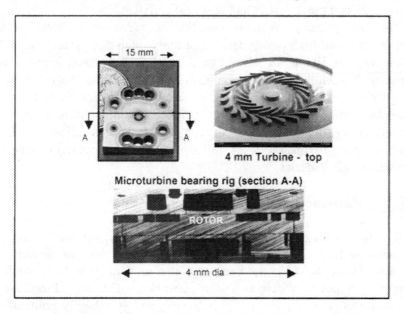

*Figure –1.6.* Micro-turbine development device, which consists of a 4 mm diameter single crystal silicon rotor enclosed in a stack of five bonded wafers used for micro air bearing development. Reproduced with permission from Luc Frechette at Columbia University.

Many other research groups have subsequently undertaken the development of micro-heat engines. Various approaches have been taken including micro gas turbine engines (Isomura *et al* 2002), Rankine steam turbines (Lee *et al* 2003), rotary Wankel internal combustion engine (Fu *et al* 2001), free and spring loaded piston internal combustion engines (Matta *et al* 2001, Toriyama *et al* 2003), and thermal-expansion-actuated piezoelectric power generators (Whalen *et al* 2003, Santavicca *et al* 2003).

Most of these and similar efforts are at initial stages of development and predicted performance has not been demonstrated. However, predictions range from 0.1-10W of electrical power output, with typical masses ~1-5 g and volumes ~1 cm³. Microengines are not expected to reduce further in size due to manufacturing and efficiency constraints. At small scales, viscous drag on moving parts and heat transfer to the ambient and between components increase, which adversely impacts efficiency.

Given the relatively large power outputs of microengines (approximately 1 to 10 watts), the engine would need to operate at very low duty cycles, and secondary, short term energy storage would be needed for wireless sensor network applications. In this scenario, the engine would intermittently charge up a secondary battery or capacitor. While the low duty cycle operation would alleviate lifetime issues for the engine, it would also increase the complexity, size, and cost of the system.

The two primary potential benefits of microengines over primary batteries are their high power density and their high energy density. While high power density is an advantage in some applications, it could be a detriment for wireless sensor applications. This leaves only the higher energy density of the fuel as a significant benefit for wireless sensor networks. Future projected thermal efficiencies are approximately 20%. At this efficiency, the energy density of the system is below 6 kJ/cm³, constituting approximately a 2X improvement over the energy density of primary lithium batteries.

## 1.6     Radioactive power sources

Radioactive materials contain extremely high energy densities. As with hydrocarbon fuels, this energy has been used on a much larger scale for decades. However, it has not been exploited on a small scale as would be necessary to power wireless sensor networks. The use of radioactive materials can pose a serious health hazard, and is a highly political and controversial topic. It should, therefore, be noted that the goal here is neither to promote nor discourage investigation into radioactive power sources, but to present their potential, and the research being done in the area.

The total energy emitted by radioactive decay of a material can be expressed as in equation 1.1.

$$E_t = A_c E_e T \qquad\qquad\qquad (1.1)$$

where $E_t$ is the total emitted energy, $A_c$ is the activity, $E_e$ is the average energy of emitted particles, and $T$ is the time period over which power is collected. Table 1.3 lists several potential radioisotopes, their half-lives, specific activities, and energy densities based on radioactive decay. It should be noted that materials with lower activities and higher half-lives will produce lower power levels for more time than materials with comparatively short half-lives and high specific activities. The half-life of the material has been used as the time over which power would be collected. Only alpha and beta emitters have been included because of the heavy shielding needed for gamma emitters. Finally, uranium 238 is included for purposes of comparison only.

*Table –1.3.* Comparison of radio-isotopes.

| Material | Half-life (years) | Activity volume density (Ci/cm$^3$) | Energy Density (J/cm$^3$) |
|---|---|---|---|
| $^{238}$U | 4.5 X 10$^9$ | 6.34 X 10$^{-6}$ | 2.23 X 10$^{10}$ |
| $^{63}$Ni | 100.2 | 506 | 1.6 X 10$^8$ |
| $^{32}$Si | 172.1 | 151 | 3.3 X 10$^8$ |
| $^{90}$Sr | 28.8 | 350 | 3.7 X 10$^8$ |
| $^{32}$P | 0.04 | 5.2 X 10$^5$ | 2.7 X 10$^9$ |

While the energy density numbers reported for radioactive materials are extremely attractive, it must be remembered that in most cases the energy is being emitted over a very long period of time. Second, efficient methods of converting this power to electricity at small scales do not exist. Therefore, efficiencies would likely be extremely low.

Recently, Li and Lal (Li and Lal 2002) have used the $^{63}$Ni isotope to actuate a conductive cantilever. As the beta particles (electrons) emitted from the $^{63}$Ni isotope collect on the conductive cantilever, there is an electrostatic attraction. At some point, the cantilever contacts the radioisotope and discharges, causing the cantilever to oscillate. Up to this point the research has only demonstrated the actuation of a cantilever, and not electric power generation. However, electric power could be generated from an oscillating cantilever. The reported power output, defined as the change over time in the combined mechanical and electrostatic energy stored in the cantilever, is 0.4 pW from a 4mm X 4mm thinfilm of $^{63}$Ni. This power level is equivalent to 0.52 µW/cm$^3$. However, it should be noted that using 1 cm$^3$ of $^{63}$Ni is impractical. The reported efficiency of the device is 4 X 10$^{-6}$.

## 2. POWER DISTRIBUTION

In addition to storing power on a wireless node, in certain circumstances power can be distributed to the node from a nearby energy rich source. It is difficult to characterize the effectiveness of power distribution methods by the same metrics (power or energy density) because in most cases the power received at the node is more a function on how much power is transferred rather than the size of the power receiver at the node. Nevertheless an effort is made to characterize the effectiveness of several power distribution methods as they apply to wireless sensor networks.

## 2.1     Electromagnetic (RF) power distribution

The most common method (other than wires) of distributing power to embedded electronics is through the use of RF (Radio Frequency) radiation. Many passive electronic devices, such as electronic ID tags and smart cards, are powered by a nearby energy rich source that transmits RF energy to the passive device. The device then uses that energy to run its electronics (Friedman *et al* 1997, Hitachi 2003). This solution works well, as evidenced by the wide variety of applications where it is used, if there is a high power scanner or other source in very near proximity to the wireless device. It is, however, less effective in dense ad-hoc networks where a large area must be flooded with RF radiation to power many wireless sensor nodes.

Using a very simple model neglecting any reflections or interference, the power received by a wireless node can be expressed by equation 1.2 (Smith 1998).

$$P_r = \frac{P_0 \lambda^2}{4\pi R^2} \qquad\qquad (1.2)$$

where $P_o$ is the transmitted power, $\lambda$ is the wavelength of the signal, and $R$ is the distance between transmitter and receiver. Assume that the maximum distance between the power transmitter and any sensor node is 5 meters, and that the power is being transmitted to the nodes in the 2.4 – 2.485 GHz frequency band, which is the unlicensed industrial, scientific, and medical band in the United States. Federal regulations limit ceiling mounted transmitters in this band to 1 watt or lower. Given a 1 watt transmitter, and a 5 meter maximum distance the power received at the node would be 50 µW, which is probably on the borderline of being really useful. However, in reality the power transmitted will fall off at a rate faster than $1/R^2$ in an indoor environment. A more likely figure is $1/R^4$. While the 1 watt limit on a transmitter is by no means general for indoor use, it is usually the case that some sort of safety limitation would need to be exceeded in order to flood a room or other area with enough RF radiation to power a dense network of wireless devices.

## 2.2    Wires, acoustic, light, etc.

Other means of transmitting power to wireless sensor nodes might include wires, acoustic emitters, and light or lasers. Distributing power through a wired power grid may be effective in certain circumstances. For example, if a new "smart" building was being designed, then wires for distributed sensors could be included in the design. However, in most situations, sensors would be distributed in an existing environment and data communication would take place wirelessly. Installing wires for power distribution and/or data communication would be cost prohibitive in most situations. Energy in the form of acoustic waves has a far lower power density than is sometimes assumed. A sound wave of 100 dB in sound level only has a power level of 0.96 $\mu$W/cm$^2$. One could also imagine using a laser or other focused light source to direct power to each of the nodes in the sensor network. However, to do this in a controlled way, distributing light energy directly to each node, rather than just flooding the space with light, would likely be too complex and not cost effective. If a whole space is already flooded with light, then this source of power becomes attractive. However, this situation has been classified as "power scavenging" and will be discussed in the following section.

## 3.    POWER SCAVENGING

Unlike power sources that are fundamentally energy reservoirs, power scavenging sources are usually characterized by their power density rather than energy density. Energy reservoirs have a characteristic energy density, and how much average *power* they can provide is then dependent on the lifetime over which they are operating. On the contrary, the *energy* provided by a power scavenging source depends on how long the source is in operation. Therefore, the primary metric for comparison of scavenged sources is power density, not energy density.

Power scavenging is perhaps the most attractive of the three options because the lifetime of the node would only be limited by failure of its own components. *However, it is also potentially the most difficult method to exploit because each use environment will have different forms of ambient energy, and therefore, there is no one solution that will fit all, or even a majority, of applications.*

## 3.1      Photovoltaics (solar cells)

At midday on a sunny day, the incident light on the earth's surface has a power density of roughly 100 mW/cm³. Single crystal silicon solar cells exhibit efficiencies of 15% - 20% (Randall 2003) under high light conditions, as one would find outdoors. Common indoor lighting conditions have far lower power density than outdoor light. Common office lighting provides about 100 μW/cm² at the surface of a desk. Single crystal silicon solar cells are better suited to high light conditions and the spectrum of light available outdoors (Randall 2003). Thin film amorphous silicon or cadmium telluride cells offer better efficiency indoors because their spectral response more closely matches that of artificial indoor light. Still, these thin film cells only offer about 10% efficiency. Therefore, the power available from photovoltaics ranges from about 15 mW/cm² outdoors to 10 μW/cm³ indoors. Table 1.4 shows the measured power outputs from a cadmium telluride solar cell (Panasonic BP-243318) at varying distances from a 60 watt incandescent bulb.

*Table –1.4.* Power from a cadmium telluride solar cell at various distances from a 60 watt incandescent bulb and under standard office lighting conditions.

| Distance | 20cm | 30cm | 45 cm | Office Light |
|---|---|---|---|---|
| Power (μW/cm²) | 503 | 236 | 111 | 7.2 |

A single solar cell has an open circuit voltage of about 0.6 volts. Individual cells are easily placed in series, especially in the case of thin film cells, to get almost any desired voltage needed. A current vs. voltage (I-V) curve for a typical five cell array (wired in series) is shown below in Figure 1.7. Unlike the voltage, current densities are directly dependent on the light intensity.

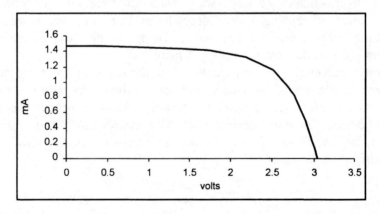

*Figure –1.7.* Typical I-V curve from a cadmium telluride solar array (Panasonic BP-243318).

Solar cells provide a fairly stable DC voltage through much of their operating space. Therefore, they can be used to directly power electronics in cases where the current load is such that it allows the cell to operate on high voltage side of the "knee" in the I-V curve and where the electronics can tolerate some deviation in source voltage. More commonly solar cells are used to charge a secondary battery. Solar cells can be connected directly to rechargeable batteries through a simple series diode to prevent the battery from discharging through the solar cell. This extremely simple circuit does not ensure that the solar cell will be operating at its optimal point, and so power production will be lower than the maximum possible. Secondly, rechargeable batteries will have a longer lifetime if a more controlled charging profile is employed. However, controlling the charging profile and the operating point of the solar cell both require more electronics, which use power themselves. An analysis needs to be done for each individual application to determine what level of power electronics would provide the highest net level of power to the load electronics. Longevity of the battery is another consideration to be considered in this analysis.

## 3.2    Temperature gradients

Naturally occurring temperature variations can also provide a means by which energy can be scavenged from the environment. The maximum efficiency of power conversion from a temperature difference is equal to the Carnot efficiency, which is given as equation 1.3.

$$\eta = \frac{T_{high} - T_{low}}{T_{high}} \tag{1.3}$$

Assuming a room temperature of 20 °C, the efficiency is 1.6% from a source 5 °C above room temperature and 3.3% for a source 10 °C above room temperature.

A reasonable estimate of the maximum amount of power available can be made assuming heat conduction through silicon material. Convection and radiation would be quite small compared to conduction at small scales and low temperature differentials. The amount of heat flow (power) is given by equation 1.4.

$$q' = k \frac{\Delta T}{L} \tag{1.4}$$

where $k$ is the thermal conductivity of the material and $L$ is the length of the material through which the heat is flowing. The conductivity of silicon is approximately 140 W/mK. Assuming a 5 °C temperature differential and a length of 1 cm, the heat flow is 7 W/cm$^2$. If Carnot efficiency could be obtained, the resulting power output would be 117 mW/cm$^2$. While this is an excellent result compared with other power sources, one must realize demonstrated efficiencies are well below the Carnot efficiency.

A number of researchers have developed systems to convert power from temperature differentials to electricity. The most common method is through thermoelectric generators that exploit the Seebeck effect to generate power. For example Stordeur and Stark (Stordeur and Stark 1997) have demonstrated a micro-thermoelectric generator capable of generating 15 μW/cm$^2$ from a 10 °C temperature differential. Furthermore, they report a technology limit of about 30 μW/cm$^2$ for the technology used. Recently Applied Digital Solutions have developed a thermoelectric generator soon to be marketed as a commercial product. The generator is reported as being able to produce 40 μW of power from a 5 °C temperature differential using a device 0.5 cm$^2$ in area and a few millimeters thick (Pescovitz 2002). The output voltage of the device is approximately 1 volt. The thermal-expansion actuated piezoelectric generator referred to earlier (Whalen *et al* 2003) has also been proposed as a method to convert power from ambient temperature gradients to electricity.

## 3.3    Human power

An average human body burns about 10.5 MJ of energy per day. (This corresponds to an average power dissipation of 121 W.) Starner has proposed tapping into some of this energy to power wearable electronics (Starner 1996). For example watches are powered using both the kinetic energy of a swinging arm and the heat flow away from the surface of the skin (Seiko 2003).

The conclusion of studies undertaken at MIT suggests that the most energy rich and most easily exploitable source occurs at the foot during heel strike and in the bending of the ball of the foot (Schenk and Paradiso 2001). This research has led to the development of piezoelectric shoe inserts capable of producing an average of 330 μW/cm$^2$ while a person is walking. The shoe inserts have been used to power a low power wireless transceiver mounted to the shoes. While this power source is of great use for wireless nodes worn on a person's foot, the problem of how to get the power from the shoe to the point of interest still remains.

The sources of power mentioned above are passive power sources in that the human doesn't need to do anything other than what he or she would normally do to generate power. There is also a class of power generators that could be classified as active human power in that they require the human to perform an action that he or she would not normally perform. For example Freeplay (Freeplay 2003) markets a line of products that are powered by a constant force spring that the user must wind up. While these types of products are extremely useful, they are not very applicable to the concept of ambient intelligence or wireless sensor networks because it would be impractical and not cost efficient to individually wind up every node in a dense network.

## 3.4     Wind / air flow

Wind power has been used on a large scale as a power source for centuries. Large windmills are still common today. However, the authors' are unaware of any efforts to try to generate power at a very small scale (on the order of a cubic centimeter or smaller) from air flow. The potential power from moving air is quite easily calculated as shown in equation 1.5.

$$P = \frac{1}{2}\rho A v^3 \qquad\qquad (1.5)$$

where $P$ is the power, $\rho$ is the density of air, $A$ is the cross sectional area, and $v$ is the air velocity. At standard atmospheric conditions, the density of air is approximately 1.22 kg/m$^3$. Figure 1.8 shows the power per square centimeter versus air velocity at this density.

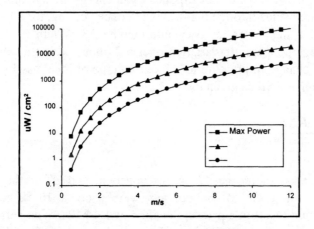

*Figure –1.8.* Maximum power density vs. air velocity. Power density assuming 20% and 5% conversion efficiencies are also shown.

Large scale windmills operate at maximum efficiencies of about 40%. Efficiency is dependent on wind velocity, and average operating efficiencies are usually about 20%. Windmills are generally designed such that maximum efficiency occurs at wind velocities around 8 m/s (or about 18 mph). At low air velocity, efficiency can be significantly lower than 20%. Figure 1.8 also shows power output assuming 20% and 5% efficiency in conversion. As can be seen from the graph, power densities from air velocity are quite promising. As there are many possible applications in which a fairly constant air flow of a few meters per second exists, it seems that research leading to the development of devices to convert air flow to electrical power at small scales is warranted.

## 3.5    Pressure variations

Variations in pressure can be used to generate power. For example one could imagine a closed volume of gas that undergoes pressure variation as the daily temperature changes. Likewise, atmospheric pressure varies throughout the day. The change in energy for a fixed volume of ideal gas due to a change in pressure is simply given by

$$\Delta E = \Delta P V \qquad\qquad\qquad (1.6)$$

where $\Delta E$ is the change in energy, $\Delta P$ is the change in pressure, and $V$ is the volume. A quick survey of atmospheric conditions around the world reveals that an average atmospheric pressure change over 24 hours is about 0.2 inches Hg or 677 Pa, which corresponds to an energy change of 677 $\mu$J/cm$^3$. If the pressure cycles through 0.2 inches Hg once per day, for a frequency of $1.16 \times 10^{-5}$ Hz, the power density would then be 7.8 nW/cm$^3$.

An average temperature variation over a 24 hour period would be about 10 °C. The change in pressure to a fixed volume of ideal gas from a 10 °C change in temperature is given by

$$\Delta P = \frac{mR\Delta T}{V} \qquad\qquad\qquad (1.7)$$

where $m$ is mass of the gas, $R$ is gas constant, and $\Delta T$ is the change in temperature. If 1 cm$^3$ of helium gas were used, a 10 °C temperature variation would result in a pressure change of 1.4 MPa. The corresponding change in energy would be 1.4 J per day, which corresponds to 17 $\mu$W/cm$^3$. While this is a simplistic analysis and assumes 100% conversion efficiency to electricity, it does give an idea of what might be theoretically expected from naturally occurring pressure variations.

To the authors' knowledge, there is no research underway to exploit naturally occurring pressure variations to generate electricity. Some clocks, such as the "Atmos clock", are powered by an enclosed volume of fluid that undergoes a phase change under normal daily temperature variations. The volume and pressure change corresponding to the phase change of the fluid mechanically actuates the clock. However, this is on a large scale, and no effort is made to convert the power to electricity.

## 3.6     Vibrations

Vibration-to-electricity conversion offers the potential for wireless sensor nodes to be self-sustaining in many environments. Low level vibrations occur in many environments including:    large commercial buildings, automobiles, aircraft, ships, trains, and industrial environments.    A combination of theory and experiment shows that about 300 $\mu W/cm^3$ could be generated from vibrations that might be found in such environments. Vibrations were measured on many surfaces inside buildings, and the resulting spectra used to calculate the amount of power that could be generated. A more detailed explanation of this process follows in Chapter 2. However, without discussing the details at this point, it does appear that conversion of vibrations to electricity can provide sufficient power for wireless sensor nodes in certain environments. Some research has been done on scavenging power from vibrations, however, it tends to be focused on a single application or technology. Therefore, a more broad look at the issue is warranted (Shearwood and Yates, 1997, Amirtharajah and Chandrakasan, 1998, Meninger *et al* 2001, Glynn-Jones *et al* 2001, Ottman *et al* 2003).

## 4.     SUMMARY OF POTENTIAL POWER SOURCES

An effort has been made to give an overview of the many potential power sources for wireless sensor networks. Well established sources, such as batteries, have been considered along with potential sources on which little or no work has been done. Because some sources are fundamentally characterized by energy density (such as batteries) while others or characterized by power density (such as solar cells) a direct comparison with a single metric is difficult. Adding to this difficulty is the fact that some power sources do not make much use of the third dimension (such as solar cells), so their fundamental metric is power per square centimeter rather than power per cubic centimeter.    Nevertheless, in an effort to compare all possible sources, a summary table is shown below as Table 1.5. Note that power density is listed as $\mu W/cm^3$, however, it is understood that in certain instances the number reported really represents $\mu W/cm^2$. Such values are marked with a "*". Note also that with only two exceptions, values listed are numbers that have been demonstrated or are based on experiments rather than theoretical optimal values. The two cases in which theoretical numbers are listed have been italicized. In many cases the theoretical best values are explained in the text above.

*Table –1.5.* Comparison of various potential power sources for wireless sensor networks. Values shown are actual demonstrated numbers except in two cases, which have been italicized.

| Power Source | $P/cm^3$ ($\mu W/cm^3$) | $E/cm^3$ ($J/cm^3$) | $P/cm^3/yr$ ($\mu W/cm^3/yr$) | Secondary storage needed | Voltage regulation needed | Commercially available |
|---|---|---|---|---|---|---|
| Primary Battery | - | 2880 | 90 | No | No | Yes |
| Secondary Battery | - | 1080 | 34 | - | No | Yes |
| Micro-fuel cell | - | 3500 | 110 | Maybe | Maybe | No |
| Ultra-capacitor | - | 50–100 | 1.6 – 3.2 | No | Yes | Yes |
| Heat engine | - | 3346 | 106 | Yes | Yes | No |
| Radioactive ($^{63}$Ni) | 0.52 | 1640 | 0.52 | Yes | Yes | No |
| Solar (outside) | 15000 * | - | - | Usually | Maybe | Yes |
| Solar (inside) | 10 * | - | - | Usually | Maybe | Yes |
| Temperature | 40 * † | - | - | Usually | Maybe | Soon |
| Human Power | 330 | - | - | Yes | Yes | No |
| Air flow | *380 ††* | - | - | Yes | Yes | No |
| Pressure Variations | *17 †††* | - | - | Yes | Yes | No |
| Vibrations | 300 | - | - | Yes | Yes | No |

* Denotes sources whose fundamental metric is power per *square* centimeter rather than power per *cubic* centimeter.
† Demonstrated from a 5 °C temperature differential.
†† Assumes air velocity of 5 m/s and 5 % conversion efficiency.
††† Based on a 1 cm$^3$ closed volume of helium undergoing a 10 °C temperature change once per day.

The following question then arises: is it preferable to use a high energy density battery that would last the lifetime of the device, or to implement an energy scavenging solution? Vibrations and solar power represent two of the more attractive solutions listed in Table 1.5. In order to give another view of the data that can perhaps better answer the question posed, Figure 1.9 shows average power available from various battery chemistries (both primary and secondary), solar power, and vibrations versus lifetime of the device being powered.

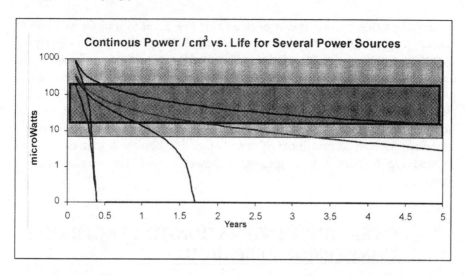

*Figure –1.9.* Power density versus lifetime for batteries, solar cells, and vibration generators.

The shaded boxes in the figure indicate the range of solar (lightly shaded) and vibration (darkly shaded) power available. Solar and vibration power output are not a function of lifetime. The reason that both solar and vibrations are shown as a box in the graph is that different environmental conditions will result in different power levels. The bottom of the box for solar power indicates the amount of power per square centimeter available in normal office lighting. The power available under direct sunlight is actually above the top of the graph shown. Likewise, the area covered by the box for vibrations covers the range of vibration sources under consideration in this study. Some of the battery traces, lithium rechargeable and zinc-air for example, exhibit an inflection point. The reason is that both battery drain and leakage are considered. For longer lifetimes, leakage becomes more dominant for some battery chemistries. The location of the inflection roughly indicates when leakage is becoming the dominant factor in reducing the amount of energy stored in the battery.

The graph indicates that if the desired lifetime of the device is in the range of 1 year or less, battery technology can provide enough energy for wireless sensor nodes (roughly 100 μW average power dissipation). However, if longer lifetimes are needed, as will usually be the case, then other options should be pursued. Also, it seems that for lifetimes of 5 years or more, a battery cannot provide the same level of power that solar cells or vibrations can provide even under poor circumstances. Therefore, battery technology will not likely meet the constraints of very many wireless sensor node applications.

Based on this survey, it was decided that low level vibrations as a power source merited a more detailed exploration. Solar power is also clearly an attractive power source for wireless sensor nodes in many environments. In most cases these are not overlapping solutions because if solar energy is present, it is likely that vibrations are not, and vice versa. Solar cells are a mature technology, and while a solar powered wireless transceiver was implemented by the authors, this is not the primary topic of the research presented here. The remainder of this chapter and book will then be devoted to studying low-level vibrations as a power source for wireless sensor networks.

# 5.   OVERVIEW OF VIBRATION-TO-ELECTRICITY CONVERSION RESEARCH

A few groups have previously devoted research effort toward the development of vibration-to-electricity converters. Yates, Williams, and Shearwood (Williams & Yates, 1995, Shearwood and Yates, 1997, Williams *et al*, 2001) have modeled and developed an electromagnetic micro-generator. The generator has a footprint of roughly 4mm X 4mm and generated a maximum of 0.3 $\mu$W from a vibration source of displacement magnitude 500 nm at 4.4 kHz. Their chief contribution, in addition to the development of the electromagnetic generator, was the development of a general second order linear model for power conversion. It turns out that this model fits electromagnetic conversion very well, and they showed close agreement between the model and experimental results. The electromagnetic generator was only 1mm thick, and thus the power density of the system was about 10 - 15 $\mu$W/cm$^3$. Interestingly, the authors do not report the output voltage and current of their device, but only the output power. This author's calculations show that the output voltage of the 0.3 $\mu$W generator would have been 8 mV, which presents a serious problem. Because the power source is an AC power source, in order to be of use by electronics it must first be rectified. In order to rectify an AC voltage source, the voltage must be larger than the forward drop of a diode, which is about 0.5 volts. So, in order to be of use, this power source would need a large linear transformer to convert the AC voltage up by at least a factor of 100 and preferably a factor of 500 to 1000, which is clearly impractical. A second issue is that the vibrations used to drive the device are of magnitude 500 nm, or 380 m/s$^2$, at 4.4 kHz. It is exceedingly difficult to find vibrations of this magnitude and frequency in many environments. These vibrations are far more energy rich than those measured in common building environments, which will be discussed at length in Chapter 2. Finally, there was no attempt in that research at either a qualitative or quantitative comparison of different methods of converting vibrations to electricity. Nevertheless the work of Yates, Williams, and Shearwood is significant in that it represents the first effort to develop micro or meso scale devices that convert vibrations to electricity (meso scale here refers to objects between the macro scale and micro scale, typically objects from a centimeter down to a few millimeters).

A second group has more recently developed an electromagnetic converter and an electrostatic converter. Several publications detail their work (Amirtharajah 1999, Amirtharajah and Chandrakasan 1998, Meninger *et al* 1999, Amirtharajah *et al* 2000, Meninger *et al* 2001). The electromagnetic converter was quite large and designed for vibrations generated by a person walking. (i.e. the person would carry the object in his/her pocket or somewhere else on the body). The device was therefore designed for a vibration magnitude of about 2 cm at about 2 Hz. (Note that these are not steady state vibrations.) Their simulations showed a maximum of 400 µW from this source under idealized circumstances (no mechanical damping or losses). While they report the measured output voltage for the device, they do not report the output power. The maximum measured output voltage was reported as 180 mV, necessitating a 10 to 1 transformer in order to rectify the voltage. The device size was 4cm X 4cm X 10cm, and if it is assumed that 400 µW of power was generated, then the power density of the device driven by a human walking would be 2.5 µW/cm$^3$. Incidentally, the researchers estimated the same power generation from a steady state vibration source driven by machine components (rotating machinery).

The electrostatic converter designed by this same group was designed for a MEMS process using a Silicon on Insulator (SOI) silicon wafer. The generator is a standard MEMS comb drive (Tang, Nguyen and Howe, 1989) except that it is used as a generator instead of an actuator. If there was an effort to consider other design topologies, the results of that effort have not been published. Amirtharajah and colleagues assume that the generator device will undergo a predetermined level of displacement, but do not show that this level of displacement is possible given a reasonable input vibration source and the dynamics of the system. In fact, this author's own calculations show that for reasonable input vibrations, and the mass of their system, the level of displacement assumed is not practical. Published simulation results for their system predict a power output of 8.6 µW for a device that is 1.5 cm X 0.5 cm X 1 mm from a vibration source at 2.52 kHz (amplitude not specified). However, to this author's knowledge, no actual test results have been published to date.

Amirtharajah and colleagues have also developed power electronics especially suited for electrostatic vibration to electricity converters for extremely low power systems. Additionally, they have developed a low power DSP (Digital Signal Processor) for sensor applications. These are both very significant achievements and contributions. In fact, perhaps it should be pointed out that the bulk of the material published about their project reports on the circuit design and implementation, not on the design and implementation of the power converter itself. The research presented in this book makes no effort to improve upon or expand their research in this area. Rather the goal of this work is to explore the design and implementation of the power converter mechanism in great detail.

Very recently a group of researchers has published material on optimal power circuitry design for piezoelectric generators (Ottman *et al* 2003, Ottman *et al* 2002). The focus of this research has been on the optimal design of the power conditioning electronics for a piezoelectric generator driven by vibrations. No effort is made to optimize the design of the piezoelectric generator itself or to design for a particular vibration source. The maximum power output reported is 18 mW. The footprint area of the piezoelectric converter is 19 $cm^2$. The height of the device is not given. Assuming a height of about 5 mm gives a power density of 1.86 $mW/cm^3$. The frequency of the driving vibrations is reported as 53.8 Hz, but the magnitude is not reported. The significant contribution of the research is a clear understanding of the issues surrounding the design of the power circuitry specifically optimized for a piezoelectric vibration to electricity converter. Again, the research presented in this work makes no effort to improve on the power electronics design of Ottman *et al*, but rather to explore the design and implementation of the power converter itself.

In order to study vibration to electricity conversion in a thorough manner, the nature of vibrations from potential sources must first be known. Chapter 2 presents the results of a study in which many commonly occurring low level vibrations were measured and characterized. A general conversion model, based on that published by Shearwood and Yates (Shearwood and Yates 1997), is also presented in chapter 2 allowing a first order prediction of potential power output of a vibration source without specifying the method of power conversion. Chapter 3 will discuss the merits of three different conversion mechanisms: electromagnetic, piezoelectric, and electrostatic. The development of piezoelectric and electrostatic converters has been pursued in detail. Chapters 4 and 5 present the modeling, design, fabrication, and test results for piezoelectric converters. The development of electrostatic converters is then presented in chapters 6, 7, and 8.

## Chapter 2

# VIBRATION SOURCES AND CONVERSION MODEL
*A Characterization of Available Vibration Sources and General Power Conversion Model*

In order to determine how much power can be converted from vibrations, the details of the particular vibration source must be considered. This chapter presents the results of a study undertaken to characterize many commonly occurring, low-level, vibrations. A general vibration to electricity model, provided by Williams and Yates (Williams and Yates 1995), is presented and discussed. The model is non-device specific, and therefore the conversion mechanism (i.e. electromagnetic, electrostatic, or piezoelectric) need not be established for the Williams and Yates model to be used. Power output can be roughly estimated given only the magnitude and frequency of input vibrations, and the overall size (and therefore mass) of the device.

## 1. TYPES OF VIBRATIONS CONSIDERED

Although conversion of vibrations to electricity is not generally applicable to all environments, it was desired to target commonly occurring vibrations in typical office buildings, manufacturing and assembly plant environments, and homes in order to maximize the potential applicability of the project. Vibrations from a range of different sources have been measured. A list of the sources measured along with the maximum acceleration magnitude of the vibrations and frequency at which that maximum occurs is shown in Table 2.1. The sources are ordered from greatest acceleration to least. It should be noted that none of the previous work cited in chapter 1 on converting vibrations to electricity has attempted to characterize a range of realistic vibration sources.

*Table –2.1.* List of vibration sources with their maximum acceleration magnitude and frequency of peak acceleration.

| Vibration Source | Peak Acc. (m/s$^2$) | Freq. of Peak (Hz) |
|---|---|---|
| Base of a 5 HP 3-axis machine tool | 10 | 70 |
| Kitchen blender casing | 6.4 | 120 |
| Clothes dryer | 3.5 | 120 |
| Door frame just after door closes | 3 | 125 |
| Small microwave oven | 2.25 | 120 |
| HVAC vents in office building | 0.2 – 1.5 | 60 |
| Wooden deck with people walking | 1.3 | 385 |
| Breadmaker | 1.03 | 120 |
| Windows (size 0.6 m X 1 m) next to a busy street | 0.7 | 100 |
| Notebook computer while CD is being read | 0.6 | 75 |
| Washing machine | 0.5 | 109 |
| Second story floor of a wood frame office building | 0.2 | 100 |
| Refrigerator | 0.1 | 240 |

## 2.     CHARACTERISTICS OF VIBRATIONS MEASURED

A few representative vibration spectra are shown in Figure 2.1. In all cases, vibrations were measured with a standard piezoelectric accelerometer. Data were acquired with a National Instruments data acquisition card at a sample rate of 20 kHz. Only the first 500 Hz of the spectra are shown because all phenomena of interest occur below that frequency. Above 500 Hz, the acceleration magnitude is essentially flat with no harmonic peaks. Measurements were taken in the same environments as the vibration sources with either the vibration source turned off (as in the case of a microwave oven) or with the accelerometer placed nearby on a surface that was not vibrating (as in the case of exterior windows) in order to ensure that vibrations signals were not the result of noise. Figure 2.2 shows measurements taken on the small microwave oven with the oven turned off and on. Note that the baseband of the signal with the microwave "off" is a factor of 10 lower than when "on". Furthermore, at the critical frequencies of 120 Hz and multiples of 120 Hz there are no peaks in acceleration when the microwave is "off".

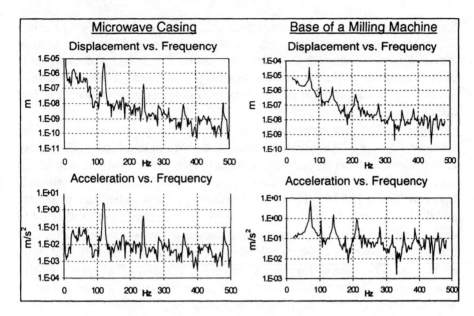

*Figure –2.1.* Two representative vibration spectra. The top graph shows displacement and the bottom graph shows acceleration.

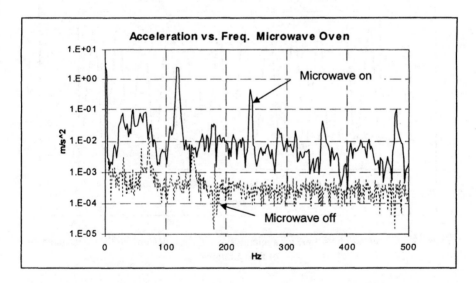

*Figure –2.2.* Acceleration taken from a microwave oven while off and on.

Vibration spectra are shown only for a small microwave oven and the base of a milling machine, however, other spectra measured but not shown here resemble the microwave and milling machine in several key respects.

First, there is a sharp peak in magnitude at a fairly low frequency with a few higher frequency harmonics. This low frequency peak will be referred to as the *fundamental vibration frequency* hereafter. The height and narrowness of the magnitude peaks are an indication that the sources are fairly sinusoidal in character, and that most of the vibration energy is concentrated at a few discrete frequencies.

Figure 2.3 shows acceleration vs. time for the microwave oven. The sinusoidal nature of the vibrations can also be seen in this figure. This sharp, low frequency peak is representative of virtually all of the vibrations measured. Second, fundamental vibration frequency for almost all sources is between 70 and 125 Hz. The two exceptions are the wooden deck at 385 Hz and the refrigerator at 240 Hz. This is significant in that it can be difficult to design very small devices to resonate at such low frequencies. Finally, note that the baseband of the acceleration spectrum is relatively flat with frequency. This means that the position spectrum falls off at approximately $1/\omega^2$ where $\omega$ is the circular frequency. Note however that the harmonic acceleration peaks are not constant with frequency. Again, this behavior is common to virtually all of the sources measured.

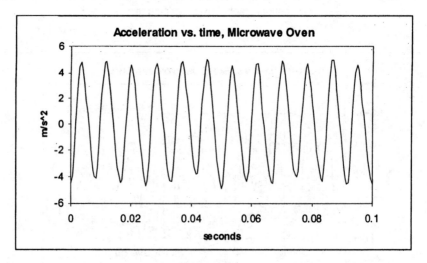

*Figure –2.3*. Acceleration vs. time for a microwave oven casing showing the sinusoidal nature of the vibrations.

Given the characteristics of the measured vibrations described above, it is reasonable to characterize a vibration source by the acceleration magnitude and frequency of the fundamental vibration mode as is done in Table 2.1. As a final note, the acceleration magnitude of vibrations measured from the casing of a small microwave oven falls about in the middle of all the sources measured. Furthermore, the frequency of the fundamental vibration mode is about 120 Hz, which is very close to that of many sources. For these reasons, the small microwave oven will be taken as a baseline when comparing different conversion techniques or different designs. When power estimates are reported, it will be assumed that a vibration source of 2.25 m/s$^2$ at 120 Hz was used unless otherwise stated.

## 3.    GENERIC VIBRATION-TO-ELECTRICITY CONVERSION MODEL

One can formulate a general model for the conversion of the kinetic energy of a vibrating mass to electrical power based on linear system theory without specifying the mechanism by which the conversion takes place. A simple model based on the schematic in Figure 2.4 has been proposed by Williams and Yates (Williams and Yates, 1995). This model is described by equation 2.1.

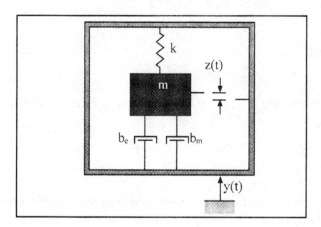

*Figure –2.4.* Schematic of a generic vibration-to-electricity converter.

$$m\ddot{z} + (b_e + b_m)\dot{z} + kz = -m\ddot{y}$$

(2.1)

where $z$ is the spring deflection, $y$ is the input displacement, $m$ is the mass, $b_e$ is the electrically induced damping coefficient, $b_m$ is the mechanical damping coefficient, and $k$ is the spring constant.

The term $b_e$ represents an electrically induced damping coefficient. The primary idea behind this model is that the conversion of energy from the oscillating mass to electricity (whatever the mechanism is that does this) looks like a linear damper to the mass spring system. This is a fairly accurate model for certain types of electro-magnetic converters like the one analyzed by Williams and Yates. For other types of converters (electrostatic and piezoelectric), this model must be changed somewhat. First, the effect of the electrical system on the mechanical system is not necessarily linear, and it is not necessarily proportional to velocity. Nevertheless, the conversion will always constitute a loss of mechanical kinetic energy, which can broadly be looked at as electrically induced "damping". Second, the mechanical damping term is not always linear and proportional to velocity. Even if this does not accurately model some types of converters, important conclusions can be made through its analysis, which can be extrapolated to electrostatic and piezoelectric systems.

The power converted to the electrical system is equal to the power removed from the mechanical system by $b_e$, the electrically induced damping. The electrically induced force is $b_e\dot{z}$. Power is simply the product of force ($F$) and velocity ($v$) if both are constants. Where they are not constants, power is given by equation 2.2.

$$P = \int_0^v F dv \qquad (2.2)$$

In the present case, $F = b_e\dot{z} = b_e v$. Equation 2.2 then becomes:

$$P = b_e \int_0^v v dv \qquad (2.3)$$

The solution of equation 2.3 is very simply $\frac{1}{2} b_e v^2$. Replacing $v$ with the equivalent $\dot{z}$ yields the expression for power in equation 2.4.

$$P = \frac{1}{2} b_e \dot{z}^2 \qquad (2.4)$$

A complete analytical expression for power can be derived by solving equation 2.1 for $\dot{z}$ and substituting into equation 2.4. Taking the Laplace transform of equation 2.1 and solving for the variable Z yields the following equation:

$$Z = \frac{-ms^2 Y}{ms^2 + (b_e + b_m)s + k} \tag{2.5}$$

where $Z$ is the Laplace transform of spring deflection, $Y$ is the Laplace transform of input displacement, and $s$ is the Laplace variable (note: $dz/dt = sZ$).

Replacing the damping coefficients $b_e$ and $b_m$ with the unitless damping ratios $\zeta_e$ and $\zeta_m$ according to the relationship $b = 2m\zeta\omega_n$, $k$ with $\omega_n^2$ according to the relationship $\omega_n^2 = k/m$, and $s$ with the equivalent $j\omega$ yields the following expression:

$$|Z| = \frac{-\omega^2}{-\omega^2 + 2(\zeta_e + \zeta_m)j\omega\omega_n + \omega_n^2}|Y| \tag{2.6}$$

where $\omega$ is the frequency of the driving vibrations, and $\omega_n$ is the natural frequency of the mass spring system.

Recalling that $|\dot{Z}| = j\omega|Z|$ and rearranging terms in equation 2.6 yields the following expression for $|\dot{Z}|$, or the magnitude of $\dot{z}$.

$$|\dot{Z}| = \frac{-j\omega\left(\dfrac{\omega}{\omega_n^2}\right)}{2\zeta_T \dfrac{j\omega}{\omega_n} + 1 - \left(\dfrac{\omega}{\omega_n}\right)^2}|Y| \tag{2.7}$$

where $\zeta_T$ is the combined damping ratio ($\zeta_T = \zeta_e + \zeta_m$).

Substituting equation 2.7 into 2.4 and rearranging terms results in an analytical expression for the output power as shown below in equation 2.8. Note that the derivation of equation 2.8 shown here depends on the assumption that the vibration source is concentrated at a single driving frequency. In other words, no broadband effects are taken into account. However, this assumption is fairly accurate given the characteristics of the vibration sources measured.

$$|P| = \frac{m\zeta_e\omega_n\omega^2\left(\dfrac{\omega}{\omega_n}\right)^3 Y^2}{\left(2\zeta_T\dfrac{\omega}{\omega_n}\right) + \left(1 - \left(\dfrac{\omega}{\omega_n}\right)^2\right)^2} \tag{2.8}$$

where $|P|$ is the magnitude of output power.

In many cases, the spectrum of the target vibrations is known beforehand. Therefore the device can be designed to resonate at the frequency of the input vibrations. If it is assumed that the resonant frequency of the spring mass system matches the input frequency, equation 2.8 can be reduced to the equivalent expressions in equations 2.9 and 2.10. Situations in which this assumption cannot be made will be discussed later in this chapter and in chapter 9 under future work.

$$|P| = \frac{m\zeta_e\omega^3 Y^2}{4\zeta_T^2} \qquad (2.9)$$

$$|P| = \frac{m\zeta_e A^2}{4\omega\zeta_T^2} \qquad (2.10)$$

where $A$ is the acceleration magnitude of input vibrations.

Equation 2.10 shows that if the acceleration magnitude of the vibration is taken to be a constant, the output power is inversely proportional to frequency. In fact, as shown previously, the acceleration is generally either constant or decreasing with frequency. Therefore, equation 2.10 is probably more useful than equation 2.9. Furthermore, the converter should be designed to resonate at the lowest fundamental frequency in the input spectrum rather than at the higher harmonics. Also note that power is optimized for $\zeta_m$ as low as possible, and $\zeta_e$ equal to $\zeta_m$. Because $\zeta_e$ is generally a function of circuit parameters, one can design in the appropriate $\zeta_e$ if $\zeta_m$ for the device is known. Finally, power is linearly proportional to mass. Therefore, the converter should have the largest proof mass that is possible while staying within the space constraints. Figure 2.5 shows the results of simulations based on this general model. The input vibrations were based on the measured vibrations from a microwave oven as described above, and the mass was limited by the requirement that the entire system stay within 1 cm$^3$ as detailed in chapter 1. These same conditions were used for all simulations and tests throughout this book; therefore all power output values can be taken to be *normalized as power per cubic centimeter*. Figure 2.5 shows power out versus electrical and mechanical damping ratio. Note that the values plotted are the logarithm of the actual simulated values. The figure shows that for a given value of $\zeta_m$, power is maximized for $\zeta_e = \zeta_m$. However, while there is a large penalty for the case where $\zeta_m$ is greater than $\zeta_e$, there is only a small penalty for $\zeta_e$ greater than $\zeta_m$. Therefore, a highly damped system will only slightly under perform a lightly damped system provided that most of the damping is electrically induced (attributable to $\zeta_e$).

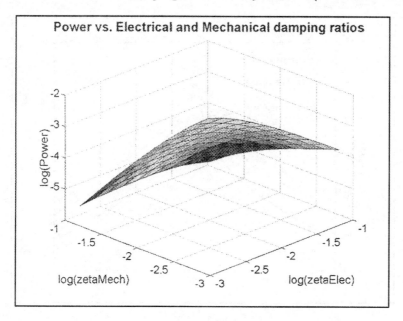

*Figure –2.5.* Simulated output power vs. mechanical and electrical damping ratios. The logarithms of the actual values are plotted.

Figure 2.5 assumes that the frequency of the driving vibrations exactly matches the natural frequency of the device. Therefore, equation 2.10 is the governing equation for power conversion. However, it is instructive to look at the penalty in terms of power output if the natural frequency of the device does not match the fundamental driving vibrations. Figure 2.6 shows the power output versus frequency assuming that the mechanical and electrically induced damping factors are equal. The natural frequency of the converter for this simulation was 100 Hz, and the frequency of the input vibrations was varied from 10 to 1000 Hz. The figure clearly shows that there is a large penalty even if there is only a small difference between the natural frequency and the frequency of the input vibrations. While a more lightly damped system has the potential for higher power output, the power output also drops off more quickly as the driving vibrations move away from the natural frequency. Based on measurements from actual devices, mechanical damping ratios of 0.01 to 0.02 are reasonable. Although, the simulation results shown in Figure 2.6 are completely intuitive, they do highlight the critical importance of designing a device to match the frequency of the driving vibrations. This should be considered a primary design consideration when designing for a sinusoidal vibration source.

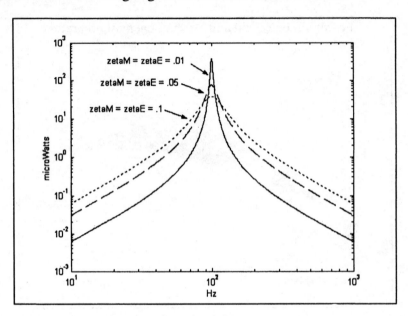

*Figure –2.6.* Power output vs. frequency for $\zeta_m$ and $\zeta_e$ equal to 0.015.

In many cases, such as HVAC ducts in buildings, appliances, manufacturing floors, etc., the frequency of the input vibrations can be measured and does not change much with time. Therefore, the converters can be designed to resonate at the proper frequency, or can have a one time adjustment done to alter their resonant frequency. However, in other cases, such as inside automobile tires or on aircraft, the frequency of input vibrations changes with time and conditions. In such cases it would be useful to actively tune the resonant frequency of the converter device. Active tuning of the device is a significant topic for future research. More will be said about this topic in chapter 9.

While the presented generic model is quite simple and neglects the details of converter implementation, it is nonetheless very useful. Because of the simplicity of the mathematics, certain functional relationships are easy to see. While the models for real converters are somewhat more complicated, the following functional relationships are nevertheless still valid.

1. The power output is proportional to the square of the acceleration magnitude of the driving vibrations.
2. Power is proportional to the proof mass of the converter, which means that scaling down the size of the converter drastically reduces potential for power conversion.
3. The equivalent electrically induced damping ratio is designable, and the power output is optimized when it is equal to the mechanical damping ratio.
4. For a given acceleration input, power output is inversely proportional to frequency. (This assumes that the magnitude of displacement is achievable since as frequency goes down, the displacement of the proof mass will increase.)
5. It is critical that the natural frequency of the conversion device closely match the fundamental vibration frequency of the driving vibrations.

## 4.     EFFICIENCY OF VIBRATION-TO-ELECTRICITY CONVERSION

The definition of the conversion efficiency is not as simple as might be expected. Generally, for an arbitrary electrical or mechanical system, the efficiency would be defined as the ratio of power output to power input. For vibration to electricity converters the power output is simple to define, however, the input power is not quite so simple. For a given vibrating mass, its instantaneous power can be defined as the product of the inertial force it exerts and its velocity. Equation 2.11 shows this relationship where $m\ddot{y}$ is the inertial force term and $\dot{y}$ is the velocity ($y$ is the displacement term).

$$P = m\ddot{y}\dot{y} \qquad (2.11)$$

The mass could be taken to be the proof mass of the conversion device. (It does not make sense to use the mass of the vibrating source, a machine tool base or large window for example, because it could be enormous. The conversion is limited by the size of the converter.) The displacement term, $y$, cannot be the displacement of the driving vibrations because the proof mass will actually undergo larger displacements than the driving vibrations. Using the displacement of the driving vibrations would therefore underestimate the input power and yield efficiencies greater than 1.

Likewise, using the theoretical displacement of the proof mass neglecting damping as the displacement term $y$ in equation 2.11 is not very useful. The displacement of the proof mass ($z$) is given by $z = Qy$ where $Q$ is the quality factor and *y* is the displacement of the input vibrations. The quality factor is the ratio of output displacement of a resonant system to input or excitation displacement. In mathematical terms, the quality factor is defined as $Q = 1/(2\zeta_T)$ for linear systems where $\zeta_T$ is the total damping ratio as described earlier. If the damping (or losses) were zero, then both the displacement of the proof mass and the force exerted by the vibration source on the converter would be infinite. The input power would also then be infinite resulting in an erroneous efficiency of zero.

The most appropriate approach is to define the input power in terms of the mechanical damping ratio, which represents pure loss. The input power would then be the product of inertial force of the proof mass and its velocity under the situation where there is no electrically induced damping. The input power is then a function of the mechanical damping ratio ($\zeta_m$). The output power is the maximum output power as defined by equation 2.10, or by more accurate models and simulations in specific cases. It is assumed that the electrically induced damping ratio ($\zeta_e$) can be arbitrarily chosen by setting circuit parameters. An efficiency curve can then be calculated that defines efficiency as a function of the mechanical damping ratio. Such a curve is shown in Figure 2.7.

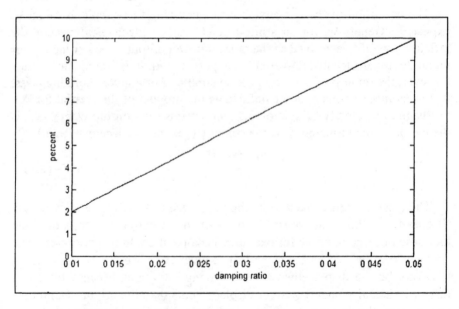

*Figure –2.7.* Conversion efficiency versus mechanical damping ratio.

The figure points out one of the difficulties in using efficiency as a metric for comparison, which is that the efficiency increases as the damping ratio increases. However, this does not mean that the output power increases with increased damping. As the mechanical damping ratio goes up, the input power goes down, and so while the ratio of output to input power increases, the actual output power decreases. This point is illustrated by Figure 2.8, which shows output power versus mechanical damping ratio. Because of this non-intuitive relationship between damping and efficiency, it is more meaningful to characterize energy conversion devices in terms of power density, defined as power per volume or $\mu W/cm^3$, rather than by efficiency. Throughout this work, devices will generally be compared by their potential power density given a standard input vibration source rather than by their efficiency. Nevertheless, for a given mechanical damping ratio, efficiency as defined and described above could be useful in comparing converters of different types.

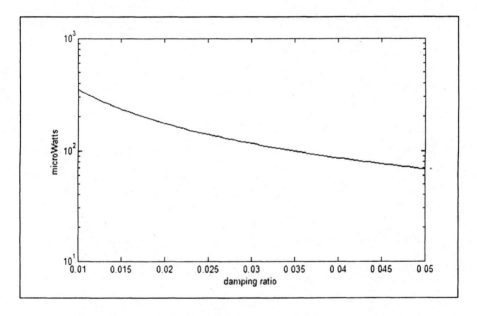

*Figure –2.8.* Simulated power output versus mechanical damping ratio.

There exists perhaps a better way to define "input" power when comparing converters of different technologies. For a given mechanical damping ratio, the "input" power could be defined as the maximum possible power conversion as defined by equation 2.10, that is the power output predicted by the technology independent model presented in this chapter. Efficiency for a given device can then be defined as the actual output power divided by the maximum possible output power for the same mechanical damping ratio. This definition of efficiency is perhaps the most useful in comparing devices from different technologies.

Chapter 3

# COMPARISON OF METHODS
*Different Methods of Converting Vibrations to Electricity*

There are three methods typically used to convert mechanical motion to an electrical signal. They are: electromagnetic (inductive), electrostatic (capacitive), and piezoelectric. These three methods are all commonly used for inertial sensors as well as for actuators. Conversion of energy intended as a power source rather than a sensor signal will use the same methods, however, the design criteria are significantly different, and therefore the suitability of each method should be re-evaluated in terms of its potential for energy conversion on the meso and micro scale. This chapter will provide an initial, primarily qualitative, comparison of these three methods. The comparison will be used as a basis to identify the areas that merit further detailed analysis.

## 1. ELECTROMAGNETIC (INDUCTIVE) POWER CONVERSION

Electromagnetic power conversion results from the relative motion of an electrical conductor in a magnetic field. Typically the conductor is wound in a coil to make an inductor. The relative motion between the coil and magnetic field cause a current to flow in the coil. A device that employs this type of conversion, taken from Amirtharajah and Chandrakasan (Amirtharajah and Chandrakasan, 1998) is shown in Figure 3.1.

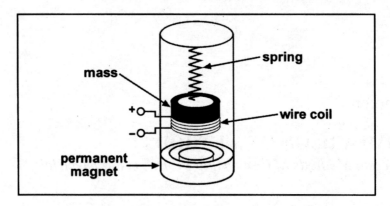

*Figure –3.1.* Electromagnetic conversion device from Amirtharajah and Chandrakasan, 1998.

The voltage on the coil is determined by Faraday's Law given in equation 3.1.

$$\varepsilon = -\frac{d\Phi_B}{dt} \qquad (3.1)$$

where $\varepsilon$ is the induced emf, and $\Phi_B$ is the magnetic flux.

In the simple case of a coil moving through a perpendicular magnetic field (as shown in Figure 3.1) of constant strength, the maximum open circuit voltage across the coil is given by equation 3.2.

$$V_{oc} = NBl\frac{dy}{dt} \qquad (3.2)$$

where $N$ is the number of turns in the coil, $B$ is the strength of the magnetic field, $l$ is the length of one coil ($2\pi r$), and $y$ is the distance the coil moves through the magnetic field.

Using the baseline vibrations of 2.25 m/s$^2$ at 120 Hz, assuming the maximum device size is 1cm$^3$, and making a few assumptions about the strength of the magnetic field and fabrication of the coil, it can easily be shown that output voltages above 100 mV are highly improbable. Table 3.1 shows open circuit voltages under various assumptions. It should be noted that the assumptions under which 124 mV could be produced are exceptionally optimistic, and frankly, it is highly unlikely that they could be achieved. In fact, the estimate of 15 to 30 mV is far more realistic given today's technology. These low voltages present a serious problem. These would be AC voltages that need to be rectified in order to be used as a power source for electronics. In order to rectify the voltages, they would have to be transformed up to the range of two to several volts necessitating a transformer with a conversion ratio on the order of 100. It would be problematic to implement such a transformer in the volume of 1cm$^3$. To do so would seriously reduce the size of the proof mass that could be designed into the system, thus reducing the potential for power conversion.

*Table –3.1*. Estimates of open circuit voltage for an inductive generator.

|  | Cond. 1 | Cond. 2 | Cond. 3 | Cond. 4 |
|---|---|---|---|---|
| Min. line and space for coil fabrication (µm) | 1 | 1 | 0.5 | 0.25 |
| Strength of magnetic field (Tesla) | 0.5 | 1 | 1 | 1 |
| Open Circuit voltage produced (mV) | 15.5 | 31 | 62 | 124 |

There are a couple of significant strengths to electromagnetic implementation. First, no separate voltage source is needed to get the process started as with electrostatic conversion. Second, the system can be easily designed without the necessity of mechanical contact between any parts, which improves reliability and reduces mechanical damping. In theory, this type of converter could be designed to have very little mechanical damping.

Two research groups (Williams *et al*, 2001, Amirtharajah and Chandrakasan, 1998) have developed electromagnetic converters. The reasons that the results of these two projects are not completely applicable to the current project have been explained in chapter 1. To summarize, the device developed by Williams *et al*, produces voltages far too low to be of use for the current project, and the device is designed to be driven by vibrations an order of magnitude higher in frequency than those under consideration. The device built and tested by Amirtharajah and Chandrakasan is significantly larger than the upper bound of 1cm$^3$ currently under consideration, and the fabrication method is not scalable down to the sizes under consideration. While Williams *et al* built and tested a microfabricated device, it is difficult to integrate this type of device with standard microelectronics. For one thing, a strong magnet has to be manually attached to the device. Additionally, just how much this magnet and its motion would affect electronics in extremely close proximity is an open question.

## 2. ELECTROSTATIC (CAPACITIVE) POWER CONVERSION

Electrostatic generation consists of two conductors separated by a dielectric (i.e. a capacitor), which move relative to one another. As the conductors move, the energy stored in the capacitor changes, thus providing the mechanism for mechanical to electrical energy conversion.

A simple rectangular parallel plate capacitor will be used to illustrate the principle of electrostatic energy conversion. The voltage across the capacitor is given by equation 3.3.

$$V = \frac{Qd}{\varepsilon_0 lw} \tag{3.3}$$

where $Q$ is charge on the capacitor, $d$ is the gap or distance between plates, $l$ is the length of the plate, $w$ is the width of the plate, and $\varepsilon_0$ is the dielectric constant of free space.

Note that the capacitance is given by $C = \varepsilon_0 lw/d$. If the charge is held constant, the voltage can be increased by reducing the capacitance, which can be accomplished either by increasing d, or reducing l or w. If the voltage is held constant, the charge can be increased by reducing d, or increasing l or w. In either case, the energy stored on the capacitor, which is given by equation 3.4, increases. An excellent discussion of charge constrained conversion versus voltage constrained conversion is given by Meninger *et al* (Meninger *et al*, 2001). The converter, then, exists of a capacitive structure, which when driven by vibrations, changes its capacitance.

$$E = \frac{1}{2}QV = \frac{1}{2}CV^2 = \frac{Q^2}{2C} \tag{3.4}$$

The primary disadvantage of electrostatic converters is that they require a separate voltage source to initiate the conversion process because the capacitor must be charged up to an initial voltage for the conversion process to start. Another disadvantage is that for many design configurations mechanical limit stops must be included to ensure that the capacitor electrodes do not come into contact and short the circuit. The resulting mechanical contact could cause reliability problems as well as increase the amount of mechanical damping.

Perhaps the most significant advantage of electrostatic converters is their potential for integration with microelectronics. Silicon micromachined electrostatic transducers are the backbone of MEMS technology. MEMS transducers use processes very similar to microelectronics. Therefore, because of the process compatibility, it is easier to integrate electrostatic converters based on MEMS technology than either electromagnetic or piezoelectric converters. Another advantage is that, unlike electromagnetic converters, appropriate voltages for microelectronics, on the order of two to several volts, can be directly generated. The modeling and design of electrostatic converters will be presented in detail in chapter 6. The fabrication and testing of electrostatic vibration based generators will be presented in chapters 7 and 8.

# 3. PIEZOELECTRIC POWER CONVERSION

Piezoelectric materials are materials that physically deform in the presence of an electric field, or conversely, produce an electrical charge when mechanically deformed. This effect is due to the spontaneous separation of charge within certain crystal structures under the right conditions producing an electric dipole. At the present time, polycrystalline ceramic is the most common piezoelectric material. Polycrystalline ceramic is composed of randomly oriented minute crystallites. Each crystallite is further divided into tiny "domains", or regions having similar dipole arrangements. Initially the polar domains are oriented randomly, resulting in a lack of macroscopic piezoelectric behavior. During manufacturing, the material is subjected to a large electrical field (on the order of 2kV/mm), which orients the polar domains in the direction of the external electrical field. The result is that the material now exhibits macroscopic piezoelectricity. If a voltage is applied in the same direction as the dipoles (the direction of the poling electric field), the material elongates in that direction. The opposite effect is also present, specifically if a mechanical strain is produced in the direction of the dipoles, a charge separation across the material (which is a dielectric) occurs, producing a voltage. A more detailed description of the piezoelectric effect is beyond the scope of this work. For a more detailed description the reader is referred to Ikeda (Ikeda 1990).

The constitutive equations for a piezoelectric material are given in equations 3.5 and 3.6.[1]

$$\delta = \sigma/Y + dE \tag{3.5}$$

$$D = \varepsilon E + d\sigma \tag{3.6}$$

where $\delta$ is mechanical strain, $\sigma$ is mechanical stress, $Y$ is the modulus of elasticity (Young's Modulus), $d$ is the piezoelectric strain coefficient, $E$ is the electric field, $D$ is the electrical displacement (charge density), $\varepsilon$ is the dielectric constant of the piezoelectric material.

Without the piezoelectric coupling term, $dE$, equation 3.5 is simply Hooke's Law. Likewise, without the coupling term, $d\sigma$, equation 3.6 is simply the dielectric equation, or a form of Gauss' law for electricity. The piezoelectric coupling provides the medium for energy conversion. The electric field across the material affects its mechanics, and the stress in the material affects its dielectric properties.

A circuit representation of a piezoelectric element is shown in Figure 3.2. The source voltage is simply defined as the open circuit voltage resulting from equation 3.6. (The open circuit condition means that the electrical displacement (D) is zero.) The expression for the open circuit voltage is given by equation 3.7.

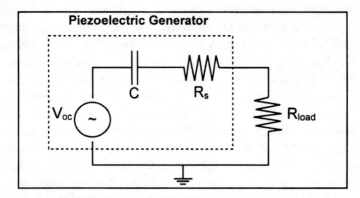

*Figure –3.2.* Circuit representation of a piezoelectric element.

$$V_{OC} = \frac{-dt}{\varepsilon}\sigma \qquad\qquad (3.7)$$

where, $t$ is the thickness of the piezoelectric material.

If the piezoelectric material undergoes a periodic or sinusoidal stress due to external vibrations, an AC open circuit voltage defined by equation 3.7 can be measured across the material. If a simple resistive load is attached to the piezoelectric generator as shown in Figure 3.2, an AC voltage ($V_{load}$) will appear across the load. The average power delivered to load is then simply $P = V_{load}^2 / 2R_{load}$. In reality, a simple resistor is not a very useful load. The voltage should be rectified and conditioned by power electronics. However, the circuit shown in Figure 3.2 gives an easy and useful calculation of power generation.

It is commonly assumed that piezoelectric devices provide high voltages and low currents (Amirtharajah 1999). However, the voltage and current levels really depend on the physical implementation and the particular electrical load circuit used. In reality, it is quite easy to design a system that produces voltages and currents in the useful range. Modeling and experiments performed by the authors show that voltages in the range of two to several volts and currents on the order of tens to hundreds of microAmps are easily obtainable. Therefore, like electrostatic converters, one of the advantages of piezoelectric conversion is the direct generation of appropriate voltages.

A second advantage is that no separate voltage source is needed to initiate the conversion process. Additionally, there is generally no need for mechanical limit stops. (There are certain cases where limit stops are important, however, this situation is uncommon.) Therefore, in principle, these devices can be designed to exhibit very little mechanical damping. Electromagnetic converters share these same advantages. It may, therefore, be said that piezoelectric converters combine most of the advantages of both electromagnetic and electrostatic converters.

The single disadvantage up to this point of piezoelectric conversion is the difficulty of implementation on the micro-scale and integration with microelectronics. While it is true that piezoelectric thin films can be integrated into MEMS processing (Lee and White, 1995), the piezoelectric coupling is greatly reduced (Verardi *et al*, 1997). Therefore, the potential for integration with microelectronics is less than that for electrostatic converters. The design and modeling of piezoelectric converters will be covered in more detail in chapter 4. chapter 5 presents test results from prototype electrostatic converters including a complete wireless sensor node powered by a piezoelectric converter.

## 4. COMPARISON OF ENERGY DENSITY OF CONVERTERS

A very useful comparison of the three methods can be made by considering the energy density inherent to each type of transducer. First consider piezoelectric transducers. If an open circuit load situation is assumed, the constitutive relationship given in equation 3.6 reduces to the expression in equation 3.8.

$$d\sigma = -\varepsilon E \qquad (3.8)$$

The energy density of a dielectric material may be expressed as $\frac{1}{2}\,\varepsilon E^2$. (Note that the units here are J/m$^3$.) Multiplying each side of equation 3.8 by $\frac{1}{2}E$ (or its equivalent $d\sigma/2\varepsilon$) yields the expression in equation 3.9. Equation 3.9 gives energy density both in terms of the electrical state of the material ($E$) and the mechanical state of the material ($\sigma$).

$$\varepsilon E^2 = \frac{\sigma^2 d^2}{\varepsilon} \tag{3.9}$$

If the yield strength of the material ($\sigma_y$) is substituted for $\sigma$, then the maximum possible energy density is given by equation 3.10.

$$E_{max} = \frac{\sigma_y^2 d^2}{2\varepsilon} \tag{3.10}$$

The piezoelectric coupling coefficient ($k$) is related to the strain coefficient ($d$) by the expression in equation 3.11. If $k$ is used rather than $d$, the expression for maximum energy density is as shown in equation 3.12.

$$d = k\sqrt{\varepsilon/Y} \tag{3.11}$$

$$E_{max} = \frac{\sigma_y^2 k^2}{2Y} \tag{3.12}$$

This may be a more intuitive form because the coupling coefficient is often used as a measure of the quality of a piezoelectric material. A coupling coefficient of 1 implies perfect coupling between the mechanical and electrical domains. Substituting in physical data for a common piezoelectric material (PZT-5H), yields a result of 35.4 mJ/cm$^3$. The more expensive and less common single crystal piezoelectric material, PZN-PT, yields 335 mJ/cm$^3$. As a practical number, assuming the properties of PZT-5H and a factor of safety of 2, the maximum energy density would be 17.7 mJ/cm$^3$.

The energy density of a capacitive device is $\frac{1}{2}\varepsilon E^2$. In the case of an electrostatic variable capacitor, the dielectric constant used is that of free space ($\varepsilon_0$). What, then, should the maximum allowable electric field for the calculation be? Maluf (Maluf 2000) arbitrarily uses 5 MV/m, which corresponds to 5 volts over a 1 $\mu$m gap. This seems too low for an estimation of maximum energy density. The maximum electric field that gas can withstand is given by Paschen's curve. At atmospheric pressure in air, the minimum voltage of Paschen's curve, which would correspond to the maximum electric field allowable, corresponds to 100 MV/m or 100 volts over a 1 μm gap. If 100 MV/m is used as a maximum electric field, the resulting energy density is 44 mJ/cm$^3$, or a little more than PZT-5H. However, if a more realistic 30 volts over a 1 μm gap were assumed (or 30 MV/m), the resulting energy density would be 4 mJ/cm$^3$.

The maximum energy density of an electromagnetic actuator (or sensor) is B$^2$/2μ$_0$ where B is the magnetic field, and μ$_0$ is the magnetic permeability. The magnetic permeability of free space is 1.26 X 10$^{-6}$ H/m. Maluf uses 0.1 Tesla as a maximum value for magnetic field, which seems quite reasonable. The resulting energy density is 4 mJ/cm$^3$. If an extremely high value of 1 Tesla were used as a maximum magnetic field, the resulting energy density would be 400 mJ/cm$^3$. Table 3.2 summarizes the maximum energy density for all three types of converters.

*Table –3.2.* Summary of maximum energy density of three types of transducers.

| Type | Governing Equation | Practical Maximum | Theoretical Max. |
|---|---|---|---|
| Piezoelectric | $u = \sigma_y^2 k^2 / 2Y$ | 17.7 mJ/cm$^3$ | 335 mJ/cm$^3$ |
| Electrostatic | $u = \dfrac{1}{2}\varepsilon E^2$ | 4 mJ/cm$^3$ | 44 mJ/cm$^3$ |
| Electromagnetic | $u = B^2 / 2\mu_0$ | 4 mJ/cm$^3$ | 400 mJ/cm$^3$ |

## 5.   SUMMARY OF CONVERSION MECHANISMS

The above discussion has been a primarily qualitative comparison of the three methods of power conversion. The purpose of performing this comparison is to serve as a basis for narrowing the range of design possibilities before performing detailed analysis, design, and optimization. The primary advantages and disadvantages of each type of converter based on this comparison are summarized in Table 3.3. It is believed that information summarized in Table 3.3 is sufficient to rule out electromagnetic converters as a suitable possibility implementation. While electromagnetic converters may be useful for larger systems, or systems exhibiting vibrations of far greater acceleration magnitude than those under consideration, they are not as suitable in the context under consideration. Piezoelectric converters exhibit all of the advantages of electromagnetic converters while additionally directly providing useful voltages and exhibiting higher practical energy densities. Furthermore, the only disadvantage of piezoelectric converters is also common to electromagnetic converters. Therefore, there is no advantage of electromagnetic over piezoelectric conversion.

*Table –3.3.* Summary of the comparison of the three conversion mechanisms.

| Type | Advantages | Disadvantages |
|---|---|---|
| Piezoelectric | No separate voltage source. Voltages of 2 to 10 volts. No mechanical stops. Highest energy density. | Microfabrication processes are not compatible with standard CMOS processes and piezo thin films have poor coupling. |
| Electrostatic | Easier to integrate with electronics and microsystems. Voltages of 2 to 10 volts. | Separate voltage source needed. Mechanical stops needed. |
| Electromagnetic | No separate voltage source. No mechanical stops. | Maximum voltage of 0.1 volts. Difficult to integrate with electronics and microsystems. |

Because piezoelectric and electrostatic converters each have unique advantages, a detailed study of these two types has been performed. The following chapters will discuss in detail the analysis, design, optimization, fabrication, and testing of piezoelectric converters (chapters 4 – 5) and electrostatic converters (chapters 6 – 8).

---

[1] Different nomenclature conventions are used in the literature when dealing with piezoelectric systems. Perhaps the most common is the convention used by Tzou (Tzou 1993) in which $T$ is used as the stress variable defined as stress induced by mechanical strain and piezoelectric effects. (Note that Tzou uses $\sigma$ as the stress induced only by mechanical strain.) $S$ is used as the strain variable, $s$ is the compliance (the inverse of elastic constant), $\varepsilon$ is the dielectric constant, $d$ is the piezoelectric strain constant, $E$ is the electric field, and $D$ is the electrical displacement. Furthermore, in the fully general case, each of these variables is a tensor or vector. While this is the convention most commonly used, it seems overly burdensome in the current context, which is quite simple in terms of mechanics. The less common convention used by Schmidt (Schmidt, 1986) is deemed more useful for the current analysis. The primary changes from the convention used by Tzou are as follows: $\delta$ is used as strain, $\sigma$ is used as stress, and $Y$ is the elastic constant (Young's modulus). Furthermore, it is assumed that the mechanics take place along a single axis, and therefore, each variable or constant is treated as a single scalar quantity rather than a tensor. The correct value of each constant, and interpretation of each variable are determined by the specifics of the device under consideration. This convention has been used in equations 3.5 – 3.7 and will be used throughout.

Chapter 4

# PIEZOELECTRIC CONVERTER DESIGN
*Modeling and Optimization*

A qualitative comparison of electrostatic, electromagnetic, and piezoelectric converters was presented in chapter 3. Chapter 4 will consider the modeling, design, and optimization of piezoelectric converters. Basic design configurations will first be discussed and evaluated. Models are then developed and validated. These models are then used as a basis for optimization.

## 1.    BASIC DESIGN CONFIGURATION

The piezoelectric constitutive equations were presented as equations 3.5 and 3.6 in the previous chapter. They will be repeated here as equations 4.1 and 4.2 for convenience.

$$\delta = \sigma/Y + dE \tag{4.1}$$

$$D = \varepsilon E + d\sigma \tag{4.2}$$

where $\delta$ is mechanical strain, $\sigma$ is mechanical stress, $Y$ is the modulus of elasticity (Young's Modulus), $d$ is the piezoelectric strain coefficient, $E$ is the electric field, $D$ is the electrical displacement (charge density), and $\varepsilon$ is the dielectric constant of the piezoelectric material.

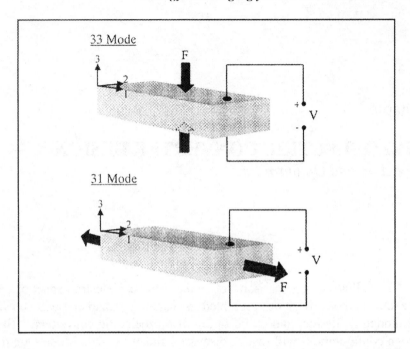

*Figure –4.1.* Illustration of 33 mode and 31 mode operation of piezoelectric material.

Figure 4.1 illustrates the two different modes in which piezoelectric material is generally used. The x, y, and z axes are labeled 1, 2, and 3. Typically, piezoelectric material is used in the 33 mode, meaning that both the voltage and stress act in the 3 direction. However, the material can also be operated in the 31 mode, meaning that the voltage acts in the 3 direction (i.e. the material is poled in the 3 direction), and the mechanical stress / strain acts in the 1 direction. Operation in 31 mode leads to the use of thin bending elements in which a large strain in the 1 direction is developed due to bending. The most common type of 31 elements are bimorphs, in which two separate sheets are bonded together, sometimes with a center shim in between them. As the element bends, the top layer of the element is in tension and bottom layer is in compression or vice versa. Therefore, if each layer is poled in the same direction and electrodes are wired properly, the current produced by each layer will add. For obvious reasons, this is termed parallel poling. Conversely, if the layers are poled in opposite directions, the voltages add. This is termed series poling. Bending elements with multiple layers (more than two), can also be made and internal electrodes provide the proper wiring between layers. In all cases, the potential for power conversion is the same. In theory, the poling and number of layers only affects the voltage to current ratio. Figure 4.2 illustrates the operation of a piezoelectric bimorph mounted as a cantilever beam and poled for series operation.

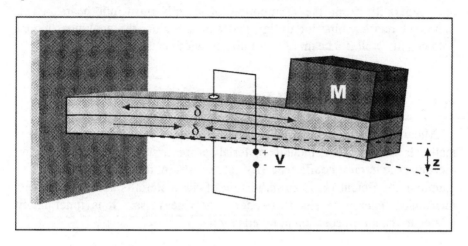

*Figure –4.2.* Operation of a piezoelectric bimorph.

Although the electrical/mechanical coupling for 31 mode is lower than for 33 mode, there is a key advantage to operating in 31 mode. The system is much more compliant, therefore larger strains can be produced with smaller input forces. Also, the resonant frequency is much lower. An immense mass would be required in order to design a piezoelectric converter operating in 33 mode with a resonant frequency somewhere around 120 Hz. Therefore, the use of bending elements operating in 31 mode is essential in this case.

A bending element could be mounted in many ways to produce a generator. A cantilever beam configuration with a mass placed on the free end (see Figure 4.2) has been chosen for two reasons. First, the cantilever mounting results in the lowest stiffness for a given size, and even with the use of bending elements it is difficult to design for operation at about 120 Hz in less than 1 cm$^3$. Second, for a given force input, the cantilever configuration results in the highest average strain for a given force input. Because the converted power is closely related to the average strain in the bender, a cantilever mounting is preferred. Note that an improvement on the simple cantilever of uniform width can be obtained varying the width of the beam. The width of the beam can be varied such that the strain along the length of the beam is the same as the strain at the fixed end, resulting in a larger average strain. This approach could result in a maximum potential average strain equal to double the average strain for the fixed width cantilever beam. For the purposes of model development, a beam of uniform width is assumed in order to keep the mathematics more manageable and because benders of uniform width are easily obtainable which makes validation of the model easier. The model developed does not lose generality from the assumption of a uniform width beam. The important relationships for design that emanate from the analytical model hold equally well if a beam of non-uniform width is used.

## 2.  MATERIAL SELECTION

Many piezoelectric materials are available. In comparing different materials a few fundamental material properties are important. The piezoelectric strain coefficient ($d$) relates strain to electric field. The coupling coefficient ($k$) is an indication of the material's ability to convert mechanical energy to electrical energy or vice versa. It is functionally related to the strain coefficient by equation 4.3.

$$k = \sqrt{\frac{Y}{\varepsilon}} d \qquad\qquad (4.3)$$

Clearly, materials with larger strain and coupling coefficients have a higher potential for energy conversion. The strain and coupling coefficients are different in 33 mode than in 31 mode, and are generally much higher in 33 mode. However, for reasons already explained, it is preferable in this case to design elements that operate in 31 mode. Two other material properties in equation 4.3 are also important. They are the dielectric constant ($\varepsilon$) and the elastic, or Young's, modulus ($Y$). A higher dielectric constant is generally preferable because it lowers the source impedance of the generator, and piezoelectric materials often have high impedance resulting in higher voltage and lower current output. The elastic modulus primarily affects the stiffness of the bender. Generally, the other material properties are more important for power conversion, and the system can be designed around the stiffness. Finally, the tensile strength of the material is very important. As mentioned earlier, the power output is related to the average strain developed. In certain cases, the design will be limited by the maximum strain that a bender can withstand. In these cases, a material with a higher tensile strength would be preferable.

*Table –4.1.* Comparison of promising piezoelectric materials.

| Property | Units | PZT | PVDF | PZN-PT |
|---|---|---|---|---|
| Stain coefficient ($d_{31}$) | $10^{-12}$ m/v | 320 | 20 | 950 |
| Strain coefficient ($d_{33}$) | $10^{-12}$ m/v | 650 | 30 | 2000 |
| Coupling coefficient ($k_{31}$) | CV/Nm | 0.44 | 0.11 | 0.5 |
| Coupling coefficient ($k_{33}$) | CV/Nm | 0.75 | 0.16 | 0.91 |
| Dielectric constant | $\varepsilon/\varepsilon_0$ | 3800 | 12 | 4500 |
| Elastic modulus | $10^{10}$ N/m$^2$ | 5.0 | 0.3 | 0.83 |
| Tensile strength | $10^7$ N/m$^2$ | 2.0 | 5.2 | 8.3 |

Table 4.1 shows a few of the most promising piezoelectric materials and their key properties (Starner 1996, Park and Shrout, 1997, Piezo Systems Inc, 1998). Strain and coupling coefficient values are given for both 33 and 31 modes. In other locations the subscripts have been omitted for simplicity, and it is assumed that the 31 mode coefficients apply. PZT (lead zirconate titanate) is probably the most commonly used piezoelectric material at the current time because of its good piezoelectric properties. PZT is a polycrystalline ceramic that, while exhibiting excellent piezoelectric coefficients, is rather brittle. There are several versions or recipes of PZT available that all have similar but slightly different properties. The specific material used here is the commonly available PSI-5H4E (Piezo Systems Inc., 1998). This same material is used for simulation, prototyping, and testing. PVDF is a piezoelectric polymer (Schmidt, 1986) that is attractive for some applications. While some of its properties are far inferior to PZT, it may be attractive in certain applications because of its higher tensile strength and lower stiffness, and because it is not brittle like ceramics. PZN-PT (Lead Zinc Niobate – Lead Titanate) is a single crystal piezoelectric material much like PZT (Park and Shrout, 1997). It has excellent properties, however, it has just become available commercially only very recently (TRS Ceramics, 2002). It is very expensive and only very small crystals can currently be produced. Because large flat elements would be needed for generators, it is not currently viable, but would be very attractive in the future. Based on this comparison of piezoelectric materials, PZT has been chosen as the primary material for further development. However, generators based on PVDF have also been modeled and optimized for comparison. As will be shown later, PZT generators are capable of higher power output than PVDF and therefore prototypes for testing have been built using PZT.

# 3.      ANALYTICAL MODEL FOR PIEZOELECTRIC GENERATORS

Assuming this basic configuration (a bender mounted as a cantilever with a mass on the end), an analytic model can be developed based on beam theory and equations 4.1 and 4.2. A convenient method of modeling piezoelectric elements is to model both the mechanical and electrical portions of the piezoelectric system as circuit elements. The piezoelectric coupling is then modeled as a transformer (Flynn and Sanders 2002). The effective number of turns ($n^*$) for the transformer is explained below using equations 4.5 and 4.6. Figure 4.3 shows the circuit model of the piezoelectric element.

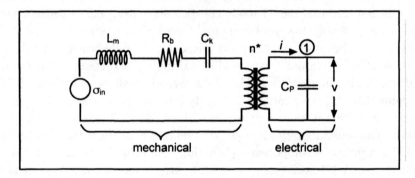

*Figure —4.3.* Circuit Representation of Piezoelectric Bimorph.

The across variable (variable acting across an element) on the electrical side is voltage ($V$) and the through variable (variable acting through an element) is current ($i$) (Rosenberg and Karnopp, 1983). The across variable on the mechanical side is stress ($\sigma$) and the through variable is strain ($\delta$). It is easier to use stress and strain as variables rather than force and tip displacement because the piezoelectric constant, $d$, relates to stress and strain. As will be shown later, strain becomes the state variable rather than the more commonly used displacement in the equations of motion. The mass attached to the end of the cantilever beam is shown as an inductor. The damper is shown as a resistor. It should be noted that the units of the coefficient $b_m$ in this model are $Ns/m^3$ rather than the conventional $Ns/m$. In other words, $b_m$ relates stress to strain rate rather than force to velocity. The stiffness term, $Y$, relating stress to strain is shown as a capacitor. $C_p$ is the capacitance of the bimorph. The vibration input is shown as a stress generator ($\sigma_{in}$), which comes from the input acceleration $\ddot{y}$. The relationship between the input vibrations ($\ddot{y}$) and an equivalent stress input is:

$$\sigma_{in} = \frac{m}{b^{**}} \ddot{y} \tag{4.4}$$

where $b^{**}$ is the geometric constant relating average bending stress to force at the beam's end.

The transformer relates stress ($\sigma$) to electric field ($E$) at zero strain, or electrical displacement ($D$) to strain ($\delta$) at zero electric field. So the equations for the transformer follow directly from equations 4.1 and 4.2, and are:

$$\sigma = -dYE \tag{4.5}$$

$$D = -dY\delta \tag{4.6}$$

The equivalent turns ratio ($n^*$) for the transformer is then $-dY$.

Once the circuit has been defined and the relationship between the physical beam and the circuit elements on the "mechanical" side of the circuit has been specified, system equations can be developed using Kirchoff's Voltage Law (KVL) and Kirchoff's Current Law (KCL). Appendix A contains a full derivation of the system equations. Only the resulting model, shown in equations 4.7 and 4.8, is presented here.

$$\ddot{\delta} = \frac{-k_{sp}}{m}\delta - \frac{b_m b^{**}}{m}\dot{\delta} + \frac{k_{sp}d}{mt_c}V + b^*\ddot{y} \tag{4.7}$$

$$\dot{V} = \frac{-Ydt_c}{\varepsilon}\dot{\delta} \tag{4.8}$$

where $V$ is the voltage at the output, $t_c$ is the thickness of a single layer of the piezoelectric material, $k_{sp}$ is the equivalent spring constant of cantilever beam, and $b^*$ is the geometric constant relating average strain to displacement at the beam's end.

Note that no electrical load has been applied to the system. The right side of Figure 4.3 is an open circuit, and so no power is actually transferred in this case. It is instructive to consider the case in which a simple resistor is used as the load. This results in the circuit model shown in Figure 4.4. The resulting change in the system equations is only minor, and is shown in equations 4.9 and 4.10.

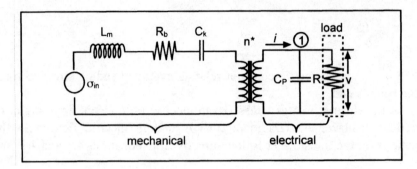

*Figure –4.4.* Circuit model of a piezoelectric bimorph with resistive load.

$$\ddot{\delta} = \frac{-k_{sp}}{m}\delta - \frac{b_m b^{**}}{m}\dot{\delta} + \frac{k_{sp}d}{mt_c}V + b^*\ddot{y} \tag{4.9}$$

$$\dot{V} = \frac{-Ydt_c}{\varepsilon}\dot{\delta} - \frac{1}{RC_p}V \tag{4.10}$$

where $R$ is the load resistance, and $C_p$ is the capacitance of the piezoelectric bender.

This model is similar in many respects to the general second order model discussed in chapter 2 and given in equation 2.1. Although this model is 3rd order, it is linear, and equation 4.9 is in the same basic form as equation 2.1. The electrical coupling term, $(k_{sp}d)/(mt_c)*V$, in equation 4.9 can be used to find the equivalent linear damping ratio, $\zeta_e$, which represents the electrically induced damping that was the basis of power conversion for the generic model of chapter 2. The equivalent electrically induced damping ratio is given by the expression in equation 4.11. (Again, see Appendix A for a full derivation of equation 4.11.)

$$\zeta_e = \frac{\omega k^2}{2\sqrt{\omega^2 + \dfrac{1}{(RC)^2}}} \tag{4.11}$$

where $k$ is the piezoelectric coupling coefficient. (See equation 4.3 for relationship between coupling coefficient ($k$) and strain coefficient ($d$)).

By proper selection of the load resistance ($R$), $\zeta_e$ will be equal to the mechanical damping ratio $\zeta$. The optimal value of $R$ can be found in two ways. The first is to simply equate the expression in equation 4.11 with the mechanical damping ratio, $\zeta$, and solve for $R$. Alternatively, if it is assumed that the frequency of the input vibrations ($\omega$) is equal to the undamped natural frequency of the device ($\omega_n$), an analytical expression for power transferred to the load can be obtained. This expression is given in equation 4.12. The optimal value of $R$ can then be found by differentiating equation 4.12 with respect to $R$, and solving for $R$. In either case, the expression in equation 4.13 is found to give the optimal load resistance. Note that if there were no piezoelectric coupling (i.e. the coupling coefficient $k = 0$), the optimal load resistance would just be $1/\omega C$, which is obvious by inspection of the circuit in Figure 4.4.

$$P = \frac{1}{\omega^2} \frac{RC^2 \left( \dfrac{Y_c dt_c b^*}{\varepsilon} \right)^2}{(4\zeta^2 + k^4)(RC\omega)^2 + 4\zeta k^2(RC\omega) + 2\zeta^2} A_{in}^2 \qquad (4.12)$$

$$R_{opt} = \frac{1}{\omega C} \frac{2\zeta}{\sqrt{4\zeta^2 + k^4}} \qquad (4.13)$$

where $A_{in}$ is the acceleration magnitude of input vibrations.

## 4. DISCUSSION OF ANALYTICAL MODEL FOR PIEZOELECTRIC GENERATORS

Because of the close similarity of the model for the piezoelectric generator to the generic conversion model presented in chapter 2, the conclusions drawn for the generic model also hold for the piezoelectric case.

The output power is proportional to the proof mass. This is not immediately obvious from the expression for power in equation 4.12. However, for a given frequency, if the mass increases other variables in equation 4.12 also change increasing the predicted power output. For example, if the mass increases, either the thickness or the width of the beam must go up to maintain the same resonant frequency. If the thickness increases, both $t_c$ and $b^*$ will increase, thus raising the predicted power out. If the width increases, $C$ increases, again resulting in a higher predicted output power.

The power output is proportional to the square of the acceleration magnitude of the driving vibrations. This follows directly from equation 4.12.

There is an equivalent electrically induced damping ratio, $\zeta_e$, and the power output is maximized when that damping ratio is equal to the mechanical damping $\zeta$ in the system. Furthermore, as shown in Figure 4.5, it is better for $\zeta_e$ to be larger than $\zeta$, rather than smaller than $\zeta$. As shown in equation 4.11, $\zeta_e$ is a function of $R$, and can therefore be controlled by the selection of the load resistance. Figure 4.5 shows the simulated power output for vs. load resistance, and Figure 4.6 shows the same simulated power output vs. the equivalent electrically induced damping factor. The mechanical damping ratio for these two simulations was 0.02. These two figures clearly demonstrate the above conclusion. In reality, the load will not be as simple as a resistor. However, the load will remove kinetic energy from the vibrating beam-mass system, and so act as electrically induced damping. Some circuit parameters can such that the power transfer to the load is maximized even with more complicated load circuitry.

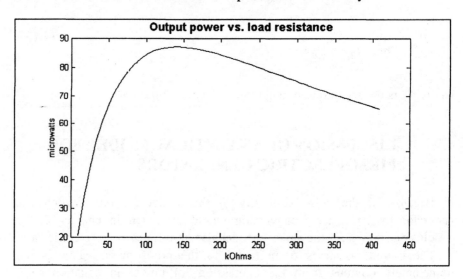

*Figure –4.5.* Simulated output power versus load resistance.

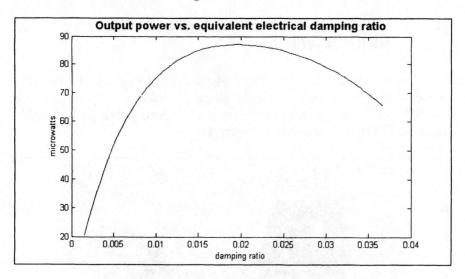

*Figure –4.6.* Simulated output power versus equivalent electrical damping ratio.

The power output is inversely related to the frequency. Again, this follows from equation 4.12. It should be noted that as the frequency of the system goes down, the displacement of the proof mass goes up. Depending on how the lower frequency was achieved (i.e. longer beam, thinner beam, increased mass, etc.) the increased displacement may be accompanied by increased strain. There is a limit to how much strain can be supported by the material, and so in some cases, the power output may be limited by the fracture strain of the piezoelectric material.

Finally, equation 4.12 assumes that the frequency of the driving vibrations is equivalent to the natural frequency of the generator device. It is critical that these two frequencies match as closely as possible. The relationship demonstrated in Figure 4.6, that the output power falls off dramatically as the driving frequency diverges from the natural frequency, also holds in the present case.

In reality it is not all that useful to design a generator to power a resistor, the value of which is chosen to optimize power transfer from the generator to the resistor. In practice the generator would be used in conjunction with a rectifier to charge up a storage capacitor, which feeds into a voltage regulator or DC-DC converter. This circuit configuration is shown later in Figure 4.14, and the implications for power conversion are discussed in more detail in section 7 of this chapter. However, at this point it should be noted that all of the above conclusions still hold for the real circuit. *The primary difference is that the equivalent electrically induced damping is no longer a function of a load resistance, but of other parameters, some of which are designable.*

# 5. INITIAL PROTOTYPE AND MODEL VERIFICATION

A bimorph made of lead zirconate titanate (PZT) with a steel center shim was used as a prototype to verify the model in equations 4.9 and 4.10. The bimorph, with attached mass (made from a relatively dense alloy of tin and bismuth) and fixture, is shown in Figure 4.7.

*Figure –4.7.* Piezoelectric (PZT) generator prototype.

The total volume of the bimorph and mass is approximately 1 cm$^3$. The converter was driven with vibrations at 120 Hz with an acceleration magnitude of 2.25 m/s$^2$. Again, these vibrations are roughly equivalent to those measured on a small microwave oven. The beam length and mass were chosen so that the system's natural frequency matched the driving frequency. The mechanical damping ratio, $\zeta$, was measured as 0.015, and the piezoelectric coupling coefficient, $k_{31}$, was measured to be 0.12. The damping ratio was measured by applying an impulse to the system, and then measuring output. An example of the resulting damped oscillations is shown in Figure 4.8. The magnitude of oscillations is measured at two separate points, $n$ periods apart. The damping ratio can then be calculated as a function of the log decrement of the two magnitudes, and the number of periods as shown in equation 4.14 (James *et al* 1994).

$$\zeta = \frac{1}{2\pi n} \ln\left(\frac{x_1}{x_2}\right) \tag{4.14}$$

where $x_1$ is the magnitude at one point of the damped oscillation, and $x_2$ is the magnitude of the damped oscillation $n$ periods later. Several measurements were taken. The mean value of these measurements was 0.014 and the standard deviation was 0.0057.

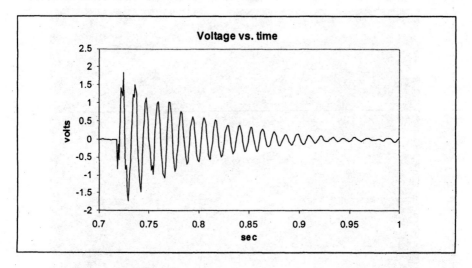

*Figure –4.8.* Damped oscillation from a force impulse to a piezoelectric generator.

The system coupling coefficient was determined by measuring the resonant frequency under open circuit ($\omega_{oc}$) and closed circuit ($\omega_{sc}$) conditions, and applying equation 4.15 (Lesieutre 1998). The system coupling coefficient ($k$) of the prototype generator was measured as 0.14. It should also be noted that the coupling coefficients of all subsequently measured prototypes were within 10% of 0.14.

$$k_{sys}^2 = \frac{\omega_{oc}^2 - \omega_{sc}^2}{\omega_{oc}^2} \tag{4.15}$$

The published coupling coefficient for the particular material used is 0.32. Because of the bonding between layers and a metal center shim, the published data from the manufacturer says to expect a working coupling coefficient of 0.75 times that of the published value (or 0.24) when using benders. However, the average measured value by the method just explained was 0.14, or just over half of what it should be. The measured value for the coupling coefficient has been used in simulations rather than the published value, which has resulted in much better agreement between the model and experiments. Other material properties were taken from published data (Piezo Systems Inc. 1998).

The prototype piezoelectric generator shown in Figure 4.7 was mounted to the vibration exciter as shown in Figure 4.9. The prototype was then driven with vibrations of 2.25 m/s$^2$ at 120 Hz.

*Figure –4.9.* First Piezoelectric prototype mounted to the vibrometer for testing.

The output was measured using a range of different load resistances. The measured and simulated output power versus load resistance is plotted in Figure 4.10. The measured and simulated voltage across the resistor is shown in Figure 4.11. The good agreement between experiments and simulations verifies that the model shown in equations 4.9 and 4.10 is sufficiently accurate to use for design and optimization purposes. Furthermore, the models can be used to obtain relatively accurate estimates of power generation.

If a bimorph poled for parallel operation was used instead, the optimal load resistance would be cut by a factor of 4, the output voltage would be cut in half, and the output current would increase by a factor of 2. In either case, the output voltage is within the right order of magnitude.

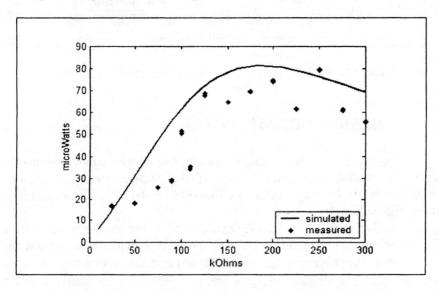

*Figure 4.10.* Measured and simulated output power versus resistive load.

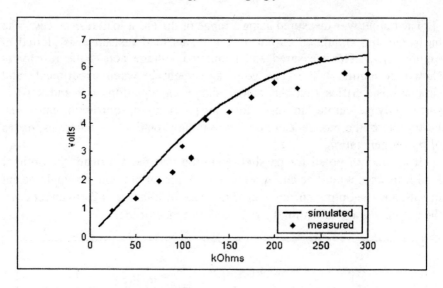

*Figure 4.11.* Measured and simulated output voltage versus resistive load.

## 6.    DESIGN OPTIMIZATION

Having defined the basic design concept, and developed and validated a model to predict the power output of that design concept, a formal mathematical optimization can be performed in order to choose dimensions and design parameters.

The objective function for the optimization is the analytical expression for power shown in equation 4.12. The output of a dynamic simulation has been used as the "objective function". However if the driving vibrations are concentrated at a single frequency and are sinusoidal in nature, then the output of the dynamic simulation matches equation 4.12 exactly. The variables over which the design can be optimized are shown in Table 4.2.

*Table –4.2.* Design variables for optimization.

| Variables | Description |
|---|---|
| $l_m$ | Length of the mass |
| $h_m$ | Height of the mass |
| $w_m$ | Width of the mass |
| $l_b$ | Length of the cantilever beam |
| $w_b$ | Width of the cantilever beam |
| $l_c$ | Length of the electrode on the beam surface |
| $t_p$ | Thickness of one piezoelectric layer |
| $t_{sh}$ | Thickness of the center shim |
| $R_{load}$ | Load resistance |

It should perhaps be noted that the piezoelectric material is not conductive. Charge can be collected from the surface and transported to the load only if the surface is covered by a conductive electrode. Portions of the surface that are not covered by an electrode act as a simple mechanical element, but do not contribute to the electrical power generation. For this reason the length of the beam and the length of the electrode are separate variables that do not necessarily have the same value.

Other parameters, such as bender capacitance and the proof mass are determined by these design variables with a few assumptions. First, it is assumed that a more dense material for the proof mass will always be preferable to a less dense material. This follows from the linear relationship between mass and power output, and the fact that the designs will generally be constrained by volume, not weight. Tungsten is the most dense commonly used material at 19 g/cm$^3$. Rhenium is actually slightly more dense, but is quite rare and extremely expensive. Other more dense materials are not commonly available and/or radioactive. Because of its extremely high hardness, tungsten is difficult to work with. Typically tungsten/nickel alloys are used which have densities around 17 g/cm$^3$. A 90% tungsten, 6% nickel, 4% copper alloy with a density of 17 g/cm$^3$ was assumed as the material for the proof mass for optimization purposes. Second, the baseline input vibrations of 2.25 m/s$^2$ at 120 Hz were used for optimization. The optimal design will, of course, be one with a resonant frequency at or very near 120 Hz. The optimization routine could be repeated for any particular vibration input magnitude and frequency, and would yield different design parameters. Third, it is assumed that only one piezoelectric layer is used on each side of the bender (this is typically referred to as a bimorph). More layers could be used; however this does not impact the output power. The overall thickness of the piezoelectric material affects the output power, and the number of layers adding up to that thickness changes the voltage to current output ratio and the optimal load resistance, but not the output power. Therefore, a single layer was assumed, and the actual design can incorporate more layers to give the appropriate voltage and current outputs. Finally, two separate optimizations were performed for two different materials, PZT and PVDF. As discussed previously, these two materials represent the most attractive commercially available materials.

The optimization problem can then be formulated as shown in Figure 4.12. There are three nonlinear constraints, and one linear constraint. The linear constraint, $l_e - l_b - l_m \leq 0$, results from the fact that it is physically impossible for the electrode length to be longer than the sum of the beam and mass lengths. The first two non-linear constraints, $(l_b + l_m)w_m h_m < 1 \text{cm}^3$ and $(l_b + l_m)w_b h_m \leq 1 \text{cm}^3$, represent overall volume constraints. As mentioned previously, the goal in this context is to design a vibration converter using a space of 1cm$^3$ or less. Finally, the maximum strain cannot exceed the yield strain of the piezoelectric material, which leads to the third non-linear constraint. The average strain is one of the state variables of the dynamic simulation, and the maximum strain can be easily calculated from the average strain and the beam geometry. Although strain is not a direct function of the design parameters, it can nevertheless be used as a non-linear optimization constraint.

Maximize: $P = f\left(l_m, h_m, w_m, l_b, w_b, l_e, t_p, t_{sh}, R_{load}\right)$

Subject to: $\left(l_b + l_m\right)w_m h_m \leq 1cm^3$

$\left(l_b + l_m\right)w_b h_m \leq 1cm^3$

$l_e - l_b - l_m \leq 0$

$l_m, h_m, w_m, l_b, w_b, l_e, t_p, t_{sh}, R_{load} \geq 0$

*Figure –4.12.* Formulation of optimization problem.

The optimization problem represented in Figure 4.12 was solved using the nonlinear constrained optimization function in the optimization toolbox in Matlab. Matlab uses the Sequential Quadratic Programming (SQP) method to solve nonlinear constrained optimization problems (Schittowski, 1985). The design that results from the optimization routine (using PZT as the piezoelectric material) is an extremely long, narrow bender with a narrow, tall proof mass. The exact design parameters that result are shown in Table 4.3.

*Table –4.3.* Optimal design parameters and predicted power output.

| Variables | Optimized Value |
|---|---|
| $l_m$ | 5 cm |
| $h_m$ | 1 cm |
| $w_m$ | 1.8 mm |
| $l_b$ | 6.3 mm |
| $w_b$ | 1.8 mm |
| $l_e$ | 6.3 mm |
| $t_p$ | 0.321 mm |
| $t_{sh}$ | 0.256 mm |
| $R_{load}$ | 463 kΩ |
| $P_{out}$ | 1.7 mW |

This design is clearly not practical. In addition to its awkward aspect ratio, it has a very high electrical impedance. Therefore practical limits need to be placed on some of the design variables. These limits can be represented as additional linear constraints. The variables to which the limits should be applied, and the precise values of the limits will depend on the application space of the converter and available materials. Table 4.4 shows one practical set of limits, the resulting design parameters, and simulated output power. Table 4.5 shows another set of limits with resulting design parameters and output power. PZT was used for the optimizations of both Table 4.4 and 4.5. Because the supplier (Piezo Systems Inc.) used by the author for PZT bimorphs only carries two thicknesses, the optimization routine was limited to 0.139 mm and 0.278 mm. Likewise, the supplier only carries elements with center shim thickness of 0.102 mm. It will be noticed that the differences between the design shown in Table 4.4 and that shown in Table 4.5 is that the total length limit is relaxed to 3 cm for the latter the thicker of the two available benders was used. Assuming that any thickness could be purchased, results in the optimal design parameters and power output shown in Table 4.6.

*Table –4.4.* Optimal design parameters and output power for one reasonable set of parameter constraints.

| Variables | Optimized Value | Range Allowed |
|-----------|-----------------|---------------|
| $l_m$ | 8.5 cm | $l_m + l_b < 1.5$ cm |
| $h_m$ | 7.7 cm | $h_m <= 7.7$ mm |
| $w_m$ | 6.7 mm | All, subject to total volume constraint |
| $l_b$ | 6.5 mm | $l_m + l_b < 1.5$ cm |
| $w_b$ | 3 mm | All, subject to total volume constraint |
| $l_c$ | 6.5 mm | All, subject to above constraint |
| $t_p$ | 0.139 mm | $t_p = 0.139$ mm |
| $t_{sh}$ | 0.102 mm | $t_{sh} = 0.1016$ |
| $R_{load}$ | 200 k$\Omega$ | All greater than zero |
| $P_{out}$ | 215 $\mu$W | |

*Table –4.5.* Optimal design parameters and output power for a second reasonable set of parameter constraints.

| Variables | Optimized Value | Range Allowed |
|-----------|-----------------|---------------|
| $l_m$ | 17.3 mm | $l_m + l_b < 3$ cm |
| $h_m$ | 7.7 mm | $h_m <= 7.7$ mm |
| $w_m$ | 3.6 mm | All, subject to total volume constraint |
| $l_b$ | 10.7 mm | $l_m + l_b < 3$ cm |
| $w_b$ | 3.2 mm | All, subject to total volume constraint |
| $l_c$ | 10.7 mm | All, subject to above constraint |
| $t_p$ | 0.278 mm | $t_p = 0.278$ mm |
| $t_{sh}$ | 0.102 mm | $t_{sh} = 0.1016$ |
| $R_{load}$ | 151 k$\Omega$ | All greater than zero |
| $P_{out}$ | 380 $\mu$W | |

*Table –4.6.* Optimal design and power output if piezo-ceramic thickness other than those available from the supplier are used

| Variables | Optimized Value | Range Allowed |
|---|---|---|
| $l_m$ | 2.56 cm | $l_m+l_b < 3$ cm |
| $h_m$ | 7.7 mm | $h_m <= 7.7$ mm |
| $w_m$ | 3.3 mm | All, subject to total volume constraint |
| $l_b$ | 4.4 mm | $l_m+l_b < 3$ cm |
| $w_b$ | 3.3 mm | All, subject to total volume constraint |
| $l_c$ | 4.4 mm | All, subject to above constraint |
| $t_p$ | 0.149 mm | All |
| $t_{sh}$ | 0.120 mm | All |
| $R_{load}$ | 170 k$\Omega$ | All greater than zero |
| $P_{out}$ | 975 $\mu$W | |

A coupling coefficient of 0.18 and a damping ratio of 0.02 were used for optimization purposes based on the measured values as described earlier. Measurements were taken with PZT-5H with a brass shim instead of the PZT-5A with a steel shim as used for the first prototype. Also, the clamp was decreased in size. These two changes account for the higher coupling coefficient (0.18 compared to 0.12 previously) and the higher damping ratios (0.02 compared to 0.014) used in the optimizations. It should be noted that all tests were performed with benders that have center shims made of metal (either steel or brass). The metal center shim adds strength to the bender and makes it much easier to cut and solder because of the brittleness of the piezoelectric ceramic. However, the presence of a center shim reduces the effective coupling coefficient (Piezo Systems Inc. 1998) by about 25%. If this published value is correct, improved power output could be obtained by using a bender without a center shim. However, the reliability of such a bender would be greatly reduced. Because of the reliability issues, it has been decided to use benders with a center shim for the purposes of this study.

A separate set of optimizations was performed using PVDF as the piezoelectric material. The published coupling coefficient of PVDF is 0.106, but the effective value in a laminate is about 75% of that number, or 0.08 (Starner 1996). However, its yield strain is far greater than that of PZT (0.02 for PVDF compared with 0.0014 for PZT). The design resulting from an unconstrained optimization is a very short, very wide design. This would naturally be expected because PVDF is so much more compliant than PZT. Again, practical limitations must be put on the design space. Tables 4.7 and 4.8 show two sets of reasonable limits on design variables with the resulting optimal designs and output parameters. The power output is somewhat lower than that for the PZT benders due to the greatly reduced coupling coefficient of PVDF. Also, the impedance of the designs is significantly higher due to the drastically reduced dielectric constant of PVDF. However, the optimization routine assumes that the bender is constructed of a single layer of piezoelectric material on each side. In reality, each side of the bender could consist of multiple layers wired in parallel, which would significantly reduce the impedance of the generator. However, it is generally the case that a PZT design will have lower impedance than a PVDF design.

*Table –4.7.* Optimal design parameters and power output for a PVDF design with one set of reasonable constraints.

| Variables | Optimized Value | Range Allowed |
|---|---|---|
| $l_m$ | 7 mm | All, subject to total volume constraint |
| $h_m$ | 7.7 mm | $h_m <= 7.7$ mm |
| $w_m$ | 10 mm | $w_m <= 10$ mm |
| $l_b$ | 3 mm | $l_b >= 3$ mm |
| $w_b$ | 10 mm | $w_b <= 10$ mm |
| $l_c$ | 3 mm | $l_c >= 3$ mm |
| $t_p$ | 0.178 mm | All |
| $t_{sh}$ | 0 mm | All |
| $R_{load}$ | 27.6 MΩ | All greater than zero |
| $P_{out}$ | 181 μW | |

*Table –4.8.* Optimal design parameters and power output for a PVDF design with a second reasonable set of constraints.

| Variables | Optimized Value | Range Allowed |
|---|---|---|
| $l_m$ | 4.6 mm | All, subject to total volume constraint |
| $h_m$ | 7.7 mm | $h_m <= 7.7$ mm |
| $w_m$ | 15 mm | $w_m <= 15$ mm |
| $l_b$ | 2.1 mm | $l_b >= 3$ mm |
| $w_b$ | 15 mm | $w_b <= 15$ mm |
| $l_c$ | 2.1 mm | $l_c >= 3$ mm |
| $t_p$ | 0.117 mm | All |
| $t_{sh}$ | 0 mm | All |
| $R_{load}$ | 23.6 MΩ | All greater than zero |
| $P_{out}$ | 211 μW | |

Although the power output is lower, there are some potential benefits to pursuing a design using PVDF. First, PVDF has greater reliability because it is not a brittle material. If the yield strength of the PVDF is exceeded, it will not crack and it will continue to function fairly well. If the yield strength of PZT is exceeded, it will crack and will lose most of its power conversion capability. Secondly, PVDF is much less expensive than PZT. So, if cost is the driving factor, and sufficient energy can be obtained with PVDF, it may be preferable.

As mentioned, the above optimizations were performed assuming a two layer bimorph (a single layer on each side of the center shim), and that the layers are arranged in series. No constraint was applied to the output voltage. In fact all of the optimal designs result in a fairly large (around 20 volts) output voltage. However, the desired output voltage can be tailored by adjusting the number of layers, while keeping the total piezo thickness constant. As mentioned previously, dividing the piezoelectric material into multiple layers wired in parallel does not affect the output power; it only affects the voltage to current ratio. Therefore, the above optimizations can easily be applied to multi-layer benders with lower output voltages.

# 7. ANALYTICAL MODEL ADJUSTED FOR A CAPACITIVE LOAD

The above analysis and optimization is based on a simple resistive load. This is not a very realistic approximation of a real electrical load. In reality, the electrical system would look something like the circuit shown in Figure 4.13. The mechanical representation for the piezoelectric bender is not shown in Figure 4.13, but is exactly the same as in Figures 3 and 4.

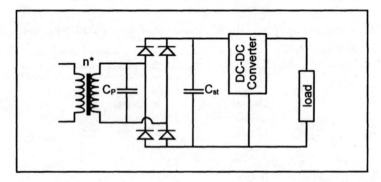

*Figure –4.13*. Piezoelectric generator with power circuitry for a piezoelectric generator. Mechanical portion of piezoelectric generator is not shown.

The piezoelectric bender charges up a large energy storage capacitor ($C_{st}$) through a full wave rectifier. The large capacitor acts as an energy buffer between the input from the piezoelectric generator the output to the integrated circuit (IC). In wireless systems, the IC will typically turn on for a short period of time and receive and transmit data, and then turn back off, or go into a sleep mode (Rabaey *et al* 2000). In the sleep mode it will dissipate very little power. In the "on" (or transmit) state, it will dissipate far more power than can be generated by the piezoelectric bender. So, during the "on" state, the voltage across the storage capacitor will fall, and during the sleep state, the voltage on the storage capacitor will increase. The IC will typically operate at very low duty cycles, around 1%. A DC-DC converter is needed so that the varying voltage across $C_{st}$ can be converted to a steady DC voltage for the IC. The system should be designed such that the average input power is at least as great as the average output power.

A large capacitor is chosen as the means of energy storage rather than a rechargeable battery for two primary reasons. First, a capacitor can be charged up by any method. In this case, it will be slowly charged up by pulses of current from the piezoelectric generator. Rechargeable batteries perform better when a specific charge-up profile is followed. While the specific charge-up profile is different for each battery chemistry, it is generally preferable to charge the battery up quickly with relatively large currents. In particular, lithium-ion batteries perform better when charged at constant current. This type of charge-up profile is simply not possible using a vibration generator unless sophisticated battery charging circuitry is used. However, the use of such circuits would greatly increase the power dissipation of the system, and therefore is not practical. The second reason is that rechargeable batteries have a relatively short shelf life. Therefore, after 1 to 2 years of operation the batteries would need to be replaced. Capacitors, on the other hand, have a virtually infinite lifetime. While it is true that batteries have a higher energy density than capacitors, the new "super" capacitors (Raible and Michel 1998, National Research Council 1997) have significantly improved energy density that is more than adequate for the current application. Rechargeable lithium-ion batteries have a maximum energy density of about 1000 $J/cm^3$. In practice, commercial batteries range from about 100 – 700 $J/cm^3$. Super capacitors have energy densities ranging from about 10 to 90 $J/cm^3$, which is about a factor of 10 lower than rechargeable batteries. However, even 5 joules of power would keep a node using an average of 100 µW alive for over 10 hours with no power input.

About 99% of the time, the IC is in sleep mode and drawing very little current, the DC-DC converter may be shut down during sleep mode as well, therefore, the vibration converter is basically just charging up the storage capacitor. A simplified circuit representation for this case is shown in Figure 4.14. This representation is useful in that it is simple enough to develop an analytical model from which design criteria may be taken.

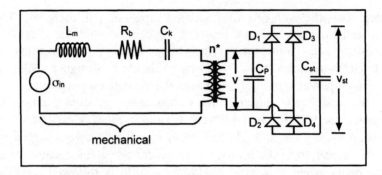

*Figure –4.14.* Simplified circuit used to analyze the charging of the storage capacitor.

There are potential three states in which the circuit shown in Figure 4.14 can operate. An ideal diode model is used in order to simplify the analysis. The assumption of ideal diodes does not change the basic functionality of the circuit. The ideal diode model assumes that the offset voltage of the diodes is zero, that the on resistance is zero, and that there is no reverse leakage. As V increases and reaches $V_{st}$, diodes D1 and D4 will turn on, diodes D2 and D3 will be off. This situation will be referred to as *stage 1*. As V decreases and reaches $-V_{st}$, then diodes D2 and D3 will conduct, and D1 and D4 will be off. This is referred to as *stage 2*. Finally if V is greater than $-V_{st}$ and less than $V_{st}$, all four diodes will be off. This is referred to as *stage 3*. Note that in stage 1, $V_{st}$ and V are equal (assuming ideal diodes), and in stage 2, $V_{st}$ is equal to $-V$. In any of the three stages, the first of the system equations is unchanged. This equation is given above as equation 4.7 and repeated here as equation 4.15.

$$\ddot{\delta} = \frac{-k_{xp}}{m}\delta - \frac{b_m b^{**}}{m}\dot{\delta} + \frac{k_{sp}d}{mt_c}V + b^*\ddot{y} \qquad (4.15)$$

In stage 3, the circuit situation is the same as that shown in Figure 4.3, and so the second of the two system equations is the same as that given above as equation 4.8 and repeated here as equation 4.16. The equivalent circuit representation for stage 1 is shown in Figure 4.15. Stage 2 results in the same circuit representation except that the polarity of V needs to be changed. The second of the two system equations for both stage 1 and stage 2 is given in equation 4.17. So, the system model is given by equations 4.15 and 4.17 for stages 1 and 2, and equations 4.15 and 4.16 for stage 3.

$$\dot{V} = \frac{-Ydt_c}{\varepsilon}\dot{\delta} \qquad (4.16)$$

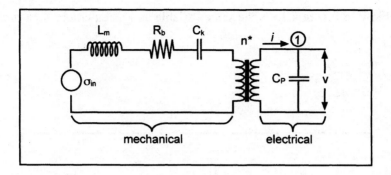

*Figure –4.15.* Equivalent circuit representation for stage 1, diodes D1 and D4 conducting.

$$\dot{V} = \frac{-YdA}{C_p + C_{st}} \dot{\delta}$$  (4.17)

where $A$ is the area covered by the electrode on the piezoelectric bender.

If a few simplifying assumptions are made, a closed form solution can be found that will provide some design intuition. In particular, it can be seen how a capacitive load circuit changes design criteria as compared to a resistive load. As with previous calculations, it is assumed that the input vibrations are a sinusoid of fixed frequency and amplitude. The second assumption that needs to be made is that the level of strain in the piezoelectric material is also a sinusoid of fixed amplitude and frequency. In other words, it is assumed that the voltage on the storage capacitor ($C_{st}$) does not affect the magnitude of the strain in the bender. This is not completely true according to the equations of motion. The voltage on $C_{st}$ affects the level of apparent damping, and therefore will affect the magnitude of the strain in the bender. However, because the apparent damping only changes a little, the affect on the strain is not dramatic. As will be shown, this assumption results in only small deviations between the closed form analytical solution, and a full dynamic simulation.

A new variable, $V_s$, can then be defined, which is the voltage that would result across the piezoelectric bender if there were no electrical load (i.e. the open circuit voltage). The circuit for this situation is shown in Figure 4.3. Following from equation 4.8, $V_s$ can then be given by equation 4.18. Given the assumption that the strain in the bender is a sinusoid of constant magnitude and frequency, a circuit representation for stage 1 (diodes D1 and D4 conducting) that is equivalent to the representation shown in Figure 4.15 is shown in Figure 4.16. *Vs* is given by $V_s(t) = V_s sin(\omega t)$. Note that the variable *V* in Figure 4.15 is the same variable as *V* is equations 4.15 through 4.17.

$$V_s = \frac{-Ydt_c}{n\varepsilon}\delta \qquad\qquad (4.18)$$

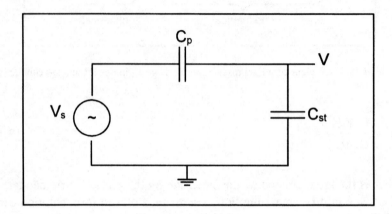

*Figure –4.16.* Equivalent circuit representation for stage 1, D1 and D4 conducting.

During each half cycle of the sinusoid a certain amount of charge is transferred to $C_{st}$, causing the voltage on $C_{st}$ to rise. This amount of charge, $\Delta Q$, is given by equation 4.19.

$$\Delta Q = \int_{t_1}^{t_2} idt = \int_{t_1}^{t_2} C_p \frac{d(V_s - V)}{dt} dt \qquad\qquad (4.19)$$

where $t_1$ is the time at which diodes D1 and D4 turn on (the point at which the circuit enters stage 1), and $t_2$ is the time at which diodes D1 and D4 turn off (the point at which the circuit leaves stage 1 and enters stage 3). This is also the point at which $V_s$ reaches its maximum point.

For simplicity, $V(t_2)$ will be referred to as $V_2$, and $V(t_1)$ as $V_1$. $\Delta Q$ can then also be given by $\Delta Q = C_{st}(V_2 - V_1)$. The increase in energy per half cycle is given by equation 4.20. Realizing the power transferred is just $2*f*\Delta E$, where $f$ is the frequency of the input vibrations, then power can be given by the expression in equation 4.21, which is a function of $V_1$ and $V_s$ rather than $V_1$ and $V_2$. Thus, all of the terms in the expression for power as shown in equation 4.21 are known at the start of each half cycle. See Appendix A for a detailed derivation of the power expression.

$$\Delta E = \frac{1}{2}(Q_2V_2 - Q_1V_1) = \frac{1}{2}C_{st}(V_2^2 - V_1^2) \tag{4.20}$$

$$P = \frac{\omega C_{st}}{2\pi(C_{st} + C_p)^2}\{C_p^2V_s^2 + 2C_{st}C_pV_sV_1 - C_pV_1^2(2C_{st} + C_p)\} \tag{4.20}$$

## 8. DISCUSSION OF ANALYTICAL MODEL CHANGES FOR CAPACITIVE LOAD

Equation 4.21 has been arranged such that $V_2$ has been replaced by $V_s$ and other constants. $V_s$ is a function of the magnitude of the input vibrations, material properties, and the geometry of the design. The only variable in equation 4.21 that will change during the operation of the generator and load circuit is $V_1$. We can see therefore that the power transfer to the storage capacitor during a given half cycle is a function of $V_1$, the voltage on the storage capacitor at the beginning of the half cycle. Figure 4.17 shows the power transferred to the storage capacitor as a function of time. The design parameters used for the simulation are those shown in Table 4.4. A storage capacitor of 1 μF was used. This compares to a piezo device capacitance of 9.4 nF. Two traces are shown. The solid line shows the result of a dynamic simulation, and the dashed line is calculated from equation 4.21. It will be noticed that the assumptions stated above do in fact alter the output power, but the agreement between the simulation and analytical solution are close enough to use the analytical solution to generate some design intuition. Figure 4.18 shows the power output as a function of $V_1$. It can clearly be seen from Figure 4.18 and less clearly from equation 4.21 that there is an optimal operating voltage. For this particular simulation, the maximum value of $V$ (the value of $V_s$) when $C_{st}$ is completely charged up is 21 volts. The optimal value of $V_1$ in this case is 10 or 11 volts depending on whether the analytical solution or the simulation is used. In general, the optimal operating point will be near half of the maximum voltage, $V_s$. In fact, if equation 4.21 is differentiated with respect to $V_1$ and solved for the optimal $V_1$, the resulting expression for the optimal value of $V_1$ is:

$$V_{1opt} = \frac{C_{st}}{2C_{st} + C_p}V_s \tag{4.22}$$

Because $C_{st}$ will naturally be much larger than $C_p$, $V_{1opt}$ is very closely equal to half $V_s$.

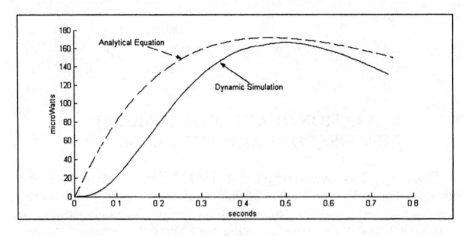

*Figure –4.17.* Simulated and analytically calculated power output versus time.  Storage capacitance is 1 μF.

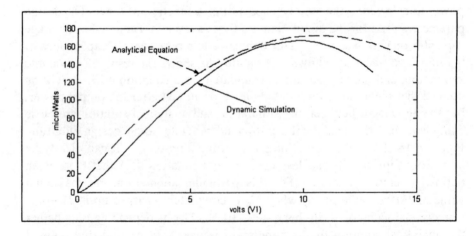

*Figure –4.18.* Simulated and analytically calculated power output versus V1, the voltage across the storage capacitor at the moment that the diodes turn on in a given half cycle. Storage capacitance is 1 μF.

A few different considerations will dictate the selection of the value of the storage capacitor. First, the capacitor has to be large enough to source the necessary current to the load when it turns on without dropping the input voltage to the DC-DC converter below an acceptable value. Second, in many instances, it will be desirable to store as much energy as possible. In this case, super capacitors that can have capacitances in excess of 1 F/cm$^3$, would seem to be a good choice. The volume constraints of the entire system should also be taken into account when selecting the capacitor. In addition to all of these considerations, equation 4.21 indicates the level of power transfer is related to value of the storage capacitance. Figure 4.19 shows the maximum power transfer as a function of the storage capacitance. The power for each value of $C_{st}$ was calculated as the maximum power transfer with respect to $V_1$, or the highest point on the graph shown in Figure 4.18. The design parameters and input for the generator are the same as those used in Figure 4.17 and 4.18. As the value of $C_p$ in this simulation is 9.4 nF, it is clear that power transfer is best when $C_{st}$ is at least many times larger than $C_p$. If $C_{st}$ is about a factor of 100 or more greater than $C_p$, the value of $C_{st}$ has very little affect on the power transfer. Therefore, $C_{st}$ should be as large as possible subject to the volume constraints of the system.

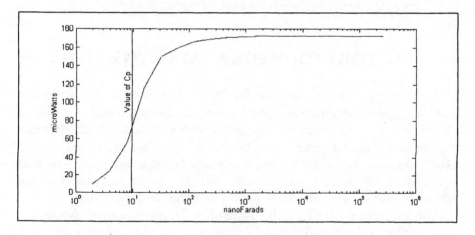

*Figure –4.19.* Power transfer to storage capacitor ($C_{st}$) as a function of the capacitance of $C_{st}$.

Clearly the capacitance of the piezo generator will also have an affect on the power transfer. This is clearly evident from equation 4.21. However, one cannot change the capacitance of the generator without changing other design parameters that will affect both the relationship between the strain and $V_s$, and the relationship between the input vibrations and average strain. Equation 4.21 indicates that all other things being constant, the power output is higher for a higher value of $C_p$ (assuming that $C_{st}$ is larger than $C_p$). However, it is difficult to see how this translates into the selection of design parameters for the bender because a change in design parameters will also affect $V_s$. A full optimization incorporating all design variables will be discussed in the following section. As will be shown, simply increasing $C_p$ by creating more, thinner layers in the bender while keeping the same overall geometry (a multilayer bender) does not increase the power output.

The model developed for a capacitive load will serve as the basis for design parameter optimization. In addition, however, it provides some engineering intuition into how the system should be designed. First, $C_{st}$ should be chosen to be as large as possible within the volume and cost constraints of the system. Also, the entire system should be designed such that the voltage on the storage capacitor during operation does not drop below about half the maximum voltage generated by the piezoelectric bender. If the voltage is allowed to drop below this value, the power transfer to the storage capacitor is significantly reduced. At the very least, this would require that the storage capacitor be large enough that during a typical transmit cycle its voltage does not drop by more than about 25% of the maximum voltage generated by the generator ($V_s$). Ideally, however, the load circuitry would adjust its duty cycle and possibly its operation in other ways depending on the voltage across the storage capacitor.

## 9. OPTIMIZATION FOR A CAPACITIVE LOAD

In the same manner as was done for the generator connected to a resistive load, a formal mathematical optimization can be performed in order to choose optimal dimensions for a generator with a capacitive load. Because the model for output power is different, the resulting model will not be exactly the same as that for the resistive load. As discussed, there is no load resistance to affect the power transfer, and the load capacitance does not affect the power transferred in the same way as the load resistance.

The variables over which the design can be optimized are exactly the same as shown above in Table 4.2 except that the load resistance ($R_{Load}$) is removed from the optimization. The storage capacitance ($C_{st}$) could be included as an optimization variable, but this seems pointless since, as was shown in the previous section, a larger $C_{st}$ will always result in more power output. However, as long as $C_{st}$ is about a factor of 100 greater than the capacitance of the bender, the value of $C_{st}$ has very little affect on the power transfer. The same assumptions regarding input vibrations, mass, the bimorph configuration, and materials (PZT and PVDF) made earlier also apply to the optimizations for a capacitive load.

Using a $C_{st}$ = 1$\mu$F, the results of a dynamic simulation of equations 4.14 – 4.16 can be used as the "objective function" for the Matlab optimization routines. The power transfer varies as a function of time, or more precisely as a function of the ratio between current and maximum voltages across the storage capacitor. It was decided that the most relevant value to use for a basis of optimization is the maximum power output (the highest point on the curve shown in Figure 4.17). As before, the only constraints on the optimization are those on the overall size, electrode length, and maximum strain.

The optimizations were performed using the material properties for PZT. The optimal design value variables and power output are shown in Table 4.9.

*Table –4.9.* Optimal design parameters and output power for capacitive load case using a storage capacitance of 1 μF.

| Variables | Optimized Value |
|---|---|
| $l_m$ | 5 cm |
| $h_m$ | 1 cm |
| $w_m$ | 1.7 mm |
| $l_b$ | 8.4 mm |
| $w_b$ | 1.7 mm |
| $l_e$ | 8.4 mm |
| $t_p$ | 0.352 mm |
| $t_{sh}$ | 0.281 mm |
| $P_{out}$ | 1.4 mW |

As was the case with the resistive load the optimization results in an impractical design. The aspect ratio is awkward, and would not likely result in a very robust structure. Interestingly, the optimal design parameters are not far removed from those for the resistive load case. Additional constraints need to be added in order for a practical design to result. The specific constraints depend on the specific application. Optimization results for two additional sets of reasonable constraints are shown in Tables 10 and 11. The constraints in Table 4.10 correspond to those shown in Table 4.4 for the resistive load case. The design was optimized such that the total length could not exceed 1.5 cm, and the thickness of the bender was constrained to that which is available from the supplier used (Piezo Systems Inc.). Table 4.11 corresponds to Table 4.6 for the resistive load case. The total length constraint is increased to 3 cm, and it is assumed that benders of any thickness could be purchased or manufactured.

*Table –4.10.* Optimal design parameters and output power for capacitive load of 1 μF incorporating one reasonable set of parameter constraints.

| Variables | Optimized Value | Range Allowed |
|---|---|---|
| $l_m$ | 7.0 mm | $l_m+l_b < 1.5$ cm |
| $h_m$ | 7.7 mm | $h_m <= 7.7$ mm |
| $w_m$ | 6.7 mm | All, subject to total volume constraint |
| $l_b$ | 8.0 mm | $l_m+l_b < 1.5$ cm |
| $w_b$ | 3 mm | All, subject to total volume constraint |
| $l_e$ | 7.7 mm | All, subject to above constraint |
| $t_p$ | 0.139 mm | $t_p = 0.139$ mm |
| $t_{sh}$ | 0.102 mm | $t_{sh} = 0.1016$ |
| $P_{out}$ | 125 μW | |

*Table –4.11.* Optimal design parameters and output power for a second reasonable set of parameter constraints. Load was 1 µF.

| Variables | Optimized Value | Range Allowed |
|-----------|-----------------|---------------|
| $l_m$ | 24.5 mm | $l_m + l_b < 3$ cm |
| $h_m$ | 7.7 mm | $h_m <= 7.7$ mm |
| $w_m$ | 3.3 mm | All, subject to total volume constraint |
| $l_b$ | 5.5 mm | $l_m + l_b < 3$ cm |
| $w_b$ | 3.3 mm | All, subject to total volume constraint |
| $l_c$ | 5.5 mm | All, subject to above constraint |
| $t_p$ | 0.149 mm | All |
| $t_{sh}$ | 0.120 mm | All |
| $P_{out}$ | 695 µW | |

The basic results for the capacitive load case closely follow the resistive load case. The optimal design variables vary somewhat between the two cases, but are not dramatically different. The power transfer is somewhat higher for a pure resistive load. Additionally, in both cases, the high cost of adding extra geometry constraints can be seen. Simply opening up the length constraint and allowing for any bender thickness increases the power output by several times for both resistive and capacitive loads.

As with the resistive load case, the optimization was performed assuming a simple layer bimorph. No constraint was placed on the voltages generated. It was mentioned before, that the bender could be designed with an appropriate number of layers to generate the desired voltage to current ratio without affecting the output power. The fact that the number of layers does not affect the power output, but only the voltage to current ratio is more intuitive for the case of the resistive load because the impedance of the load was being changed to match the impedance of the bender. However, in the current case, the load impedance is not being changed, and so one may intuitively think that using a multilayer bender with the same geometry would increase (or at least affect) the power transfer because it decreases the impedance of the bender. This, however, is not the case, as long as the storage capacitance is much greater (about a factor of 100) than the bender capacitance. The desired operating voltage across the storage capacitor can then be designed if the magnitude of the input vibrations is roughly known by specifying the number of layers in the optimal design. Figure 4.20 shows the power output versus time for the design shown in Table 4.10 for 1, 2, and 4 layer benders. Notice that the maximum power output does not change, but the voltage at which the maximum power output occurs changes. Figure 4.21 shows the voltage across the storage capacitor from the same simulation. Notice that for the benders with more layers, the voltage at which maximum power transfer occurs is reached more quickly, but that voltage is lower. A storage capacitance of 4 μF was used in these simulations.

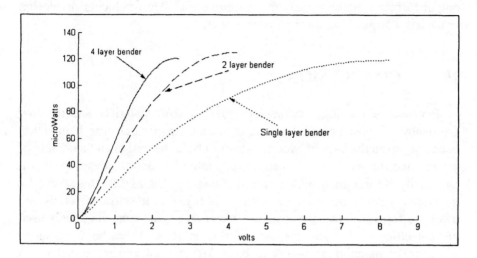

*Figure –4.20.* Power transferred versus voltage across the storage capacitor for the same design incorporating different numbers of layers in the piezoelectric bender.

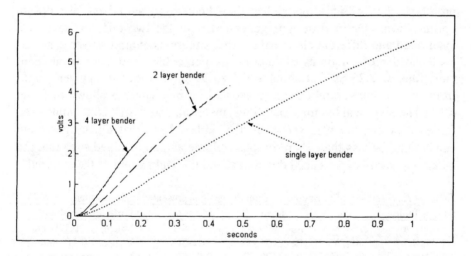

*Figure –4.21.* Voltage across storage capacitor versus time for the same design incorporating different numbers of layers in the piezoelectric bender.

As a final note regarding the operating voltage, one could of course incorporate the operating voltage as a constraint in the optimization routine. This may be necessary if only benders of a pre-specified thickness were available. Doing so, however, will further reduce the optimal power output because the design space will be further constrained. As is evident from the optimal designs shown with varying constraints, the sensitivity of the power output to many of the constraints is quite high. Thus reducing the design space has a large affect on the power output.

## 10.    CONCLUSIONS

Because of the high stiffness of piezoelectric materials and the low frequency of most potential vibration sources, a piezoelectric bender has been chosen as the basic device on which to base the design and modeling of a piezoelectric generator. A bender (or bimorph) has the advantage that it can easily be designed with lower stiffness so that higher strains can be generated with a given force input. Many piezoelectric materials are available for use. Lead zirconate titanate (PZT) is the most commonly used piezoelectric ceramic and has very good properties. It has been chosen as the primary material on which to base designs and power estimates. A piezoelectric polymer, PVDF, is also considered because of its higher yield strain and better fatigue characteristics.

A detailed model has been developed and validated with a preliminary prototype device. This model has then been used as a basis for design optimization. Optimal designs generated with the two different materials mentioned and different electrical loading situations exhibit power densities on the order of hundreds of microwatts per cubic centimeter from input vibrations of 2.25 m/s$^2$ at about 120 Hz. A summary of the power output from each of the several designs presented in the chapter is shown in Table 4.12. The electrical loading conditions, material, and design constraints used for each design are also shown in the table. Note that all designs were constrained be less than 1 cm$^3$ in total volume and constrained such that the maximum strain experienced did not exceed the yield strain of the material.

*Table –4.12.* Summary of power output from the designs presented in the chapter.

| Design Table | Power | Load | Material | Optimization Constraints |
|---|---|---|---|---|
| Table 4.4 | 215 µW | Res., 200 kΩ | PZT-5H | $l < 1.5$ cm, $t_p = 0.139$ mm |
| Table 4.5 | 380 µW | Res., 151 kΩ | PZT-5H | $l < 3$ cm, $t_p = 0.278$ mm |
| Table 4.6 | 975 µW | Res., 170 kΩ | PZT-5H | $l < 3$ cm, no constraint on $t_p$ |
| Table 4.7 | 181 µW | Res., 26.7 MΩ | PVDF | $l > 3$ mm, $w_m < 1$ cm |
| Table 4.8 | 211 µW | Res., 23.6 MΩ | PVDF | $l > 3$ mm, $w_m < 1.5$ cm |
| Table 4.10 | 125 µW | Cap., 1 µF | PZT-5H | $l < 1.5$ cm, $t_p = 0.139$ mm |
| Table 4.11 | 695 µW | Cap., 1µF | PZT-5H | $l < 3$ cm, no constraint on $t_p$ |

The power available from PZT designs considerably exceeds that available from PVDF designs (695 µW/cm$^3$ compared to 211 µW/cm$^3$).

Also, optimal designs generated with a resistive load are capable of generating a little more power than those generated with a capacitive load (975 µW/cm$^3$ compared to 695 µW/cm$^3$). However, while the model for the resistive load is useful in roughly predicting power output, validating the models, and gaining design intuition, it is not very useful in terms of practical applications. Therefore, the power output values predicted for capacitive load circuits are considered more useful and realistic.

Given the discussion presented in this chapter and previous chapters, the following simple design sequence emerges:

1. Define and design the characteristics of the load (most likely a wireless senor of some sort). Define such things as the voltage, standby current, transmit or "on" current, average power dissipation, minimum duty cycle, is duty cycle adjustable depending on energy available, etc.
2. Define the characteristics of the input vibrations. What is their average magnitude and frequency? Is the frequency and magnitude consistent over time, etc.?
3. Estimate the power potential from the vibrations using the generic power expression given in equation 2.5. Determine how much volume, or mass, is necessary to supply the power needed for the load. Is vibration conversion feasible? If so, proceed.
4. Choose or design a suitable DC-DC converter or voltage regulator for the application and full wave rectifier.
5. Choose the storage capacitor (or rechargeable battery) based on system and load constraints. These constraints will include, but are not limited to, volume, cost, maximum current draw, and maximum acceptable voltage drop during transmit or "on" state.
6. Run the optimization routine described above with the chosen storage capacitor (or rechargeable battery). Evaluate the resulting design to see if the power generation is adequate.
7. Choose the number of layers in the bender in order to get an acceptable operating voltage range as input to the DC-DC converter.
8. Evaluate bender capacitance and storage capacitance. Is storage capacitance at least a factor of 100 greater than bender capacitance? If not, re-evaluate choice of storage capacitance.

# Chapter 5

# PIEZOELECTRIC CONVERTER TEST RESULTS

A model to predict the output of piezoelectric generators was developed and discussed in the previous chapter. This model was validated with a prototype to ensure its accuracy and suitability of for use as a basis for design optimization. Optimal designs were generated and discussed. Actual converters were designed and built based on these optimizations. This chapter will discuss.the implementation of these converters and present test results showing the improvement of the optimal designs. The converters have been used to power small wireless sensor devices, and results from such tests will also be presented.

## 1. IMPLEMENTATION OF OPTIMIZED CONVERTERS

As explained in the previous chapter, the design of optimized prototypes is still constrained to materials that are commonly available. Because only a few prototypes were built, it was not feasible to ask a manufacturer to fabricate benders to our specifications. Therefore, the design was limited to benders that are available off-the-shelf. Piezo Systems Inc. carries a number of such benders that meet the specifications for this project quite well. It is quite easy to cut the benders to any size, however, the thickness of the ceramic layers is determined by what the manufacturer carries. The best commonly available bimorphs found were PZT PSI-5H4E with a brass center shim of thickness 0.1016 mm. Each of the two piezoelectric ceramic layers has a thickness of 0.1397 mm for a total bender thickness of 0.381 mm. It was also decided to use a bimorph poled for parallel operation so that the output voltages would remain in the 3 to 10 volt range.

The total volume of the device was constrained to 1cm$^3$. Because the power output is proportional to the mass of the system, a very dense material should be used for the proof mass. As explained previously, a tungsten alloy (90% tungsten, 6% nickel, 4% copper) was chosen as the material for the proof mass.

The construction of the test devices was rather simple. A base was machined from plastic. The electrodes were etched off from the bender where the mass was to be attached. Remember that, as shown in chapter 4, the optimal electrode length is generally equal to, or very close to the beam length up to the point where the mass is attached. The electrodes on the benders purchased were made of nickel and were easily etched off with common copper PC board etchant. Because the bimorphs used for these designs were poled in parallel, the center shim needed to be electrically contacted. In order to achieve this, a small slot was milled near the base of the beam (see Figure 5.1 below). The slot was just wide enough to solder a wire to the center shim. Two other wire leads were then soldered to the electrodes on either side of the bimorph. The proof mass was attached to the end of the bender using super glue. Two methods of attaching the bender to the plastic base were used: the bender was either clamped down or attached with super glue. Both methods of attachment are shown in Figure 5.1. The generator in Figure 5.1 has the dimensions given in Table 4.4 of chapter 4. Damping ratios were measured using the same procedure as outlined in Chapter 4 for the clamped beams and the beams attached with super glue. The average damping ratio for the clamped beam tests was 0.025 with a standard deviation of 0.0098, and the average damping ratio for the glued beam tests was 0.031 with a standard deviation of 0.014.

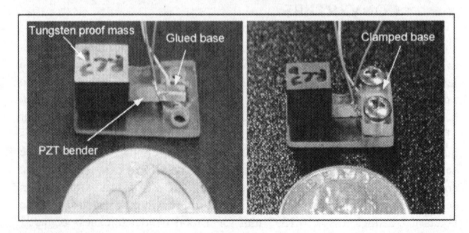

*Figure –5.1.* Piezoelectric test generators. The left picture shows the bimorph held down with super glue, the right picture shows the bimorph held down with a clamp.

A few different designs incorporating different sets of constraints were built and tested. The two designs shown previously in Tables 4 and 5 of chapter 4 were chosen as solutions incorporating a reasonable set of constraints. The primary difference between the two designs is in the geometric length constraint. For the design shown in Table 4.4 of the previous chapter, a maximum total length of 1.5 cm was used as a practical constraint in the optimization routine. This design will be referred to as *Design 1*, and is shown above in Figure 5.1. The design shown in Table 4.5 was limited to a total length of 3 cm. This design will be referred to as *Design 2*. Designs 1 and 2 are shown in Figure 5.2. In reality, the total length constraint will be determined by the specific application. The total volume constraint for each design was 1 cm$^3$, and all other constraints were the same as explained in chapter 4.

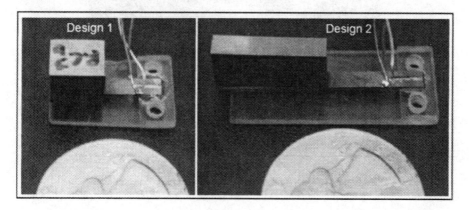

*Figure –5.2.* Two test generators built to two different sets of optimized dimensions. Design 1 is on the left and Design 2 is on the right.

## 2.    RESISTIVE LOAD TESTS

The generators were mounted on a small vibrometer as shown in Figure 5.3. The base of the generator was mounted using double sided tape. Although not a tremendously rigid attachment method, the double-sided tape has a flat frequency response within the range of interest. Because the driving vibrations of interest are low frequency (120 Hz), only frequencies up to 500 Hz were measured. The frequency response of an accelerometer mounted with tape from 0 to 500 Hz is shown in Figure 5.4.

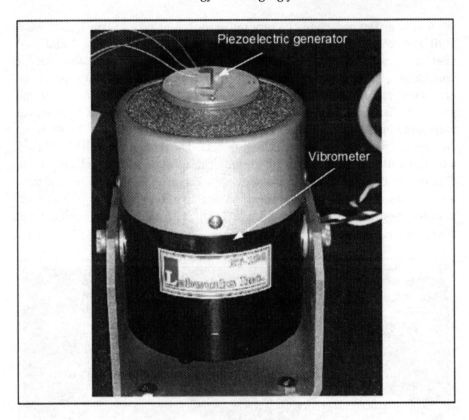

*Figure –5.3.* Vibrometer with test generator mounted.

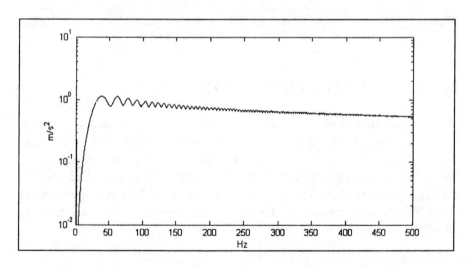

*Figure –5.4.* Frequency response of an accelerometer mounted with tape to the vibrometer. Response is flat showing that tape has no affect up to 500 Hz.

The vibrometer was calibrated before each set of tests performed. The accelerometer was used to calibrate the vibrometer outputs 0.1 volts per g (9.81 m/s$^2$). An amplifier with a gain of 10 was used with the accelerometer to output 1 volt per g. There are then 3 inputs to the amplifier that can affect the acceleration output. The vibrometer is actuated by producing a sine wave (or some other waveform) from a signal generator (Agilent 33120A), using the waveform as the input to a power amplifier (Labworks PA-138), and connecting the output of the power amplifier to the input of the vibrometer (Labworks ET-126). The magnitude and frequency of the source waveform, and the gain of the power amplifier all affect the acceleration of acceleration generated by the vibrometer. The magnitude of the source waveform was kept constant at 1 volt rms. Only the frequency and gain on the power amplifier were used to produce the desired vibrations. The vibrometer was calibrated for each set of tests done. The results of the calibration done for one set of tests are shown in Figure 5.5. The points shown are averages of three measurements taken at three different frequencies and three different gains within the range of interest. Three calibration curves are shown, one for each gain. The equations from the quadratic curve fits are shown on the figure. Note the nonlinear relationship between acceleration and frequency. Other calibrations performed, resulted in similar data.

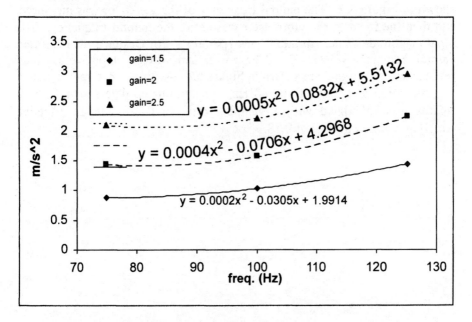

*Figure –5.5.* Acceleration versus frequency output of the vibrometer for three different power amplifier gains.

The piezoelectric generator was terminated with a resistor and the voltage across the resistor was measured. The voltage signal was first passed through a unity gain buffer to decouple the capacitance of the data acquisition system from the generator. The test circuit is shown in Figure 6. A National Instruments data acquisition card (DAQCard-AI-16XE-50) capable of acquiring 20,000 samples per second was used in conjunction with LabView software to acquire the data.

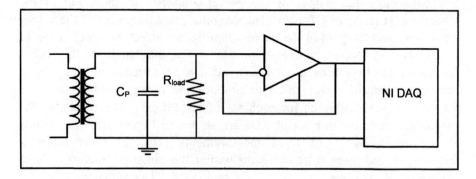

*Figure –5.6.* Measurement circuit for resistive load tests.

The output power and peak voltage versus load resistance for Design 1 is shown in Figure 5.7. The natural frequency of the generator was measured, and then the input to the vibrometer was set to the natural frequency. The input magnitude of the vibrations was 2.25 m/s$^2$. For Design 1 the measured natural frequency was 85 Hz. The power and voltage output versus load resistance for Design 2 are shown in Figure 5.8. Again the natural frequency was measured and found to be 60 Hz. The simulated data shown in both figures was calculated with a damping ratio of 0.025 and an effective coupling coefficient ($k_{31}$) of 0.18.

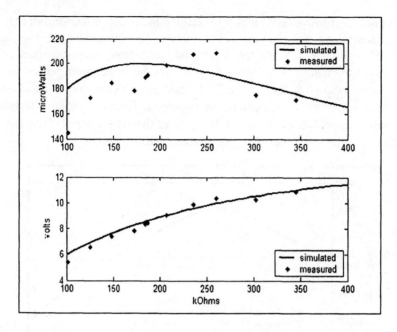

*Figure –5.7.* Measured and simulated output power and peak voltage versus load resistance for Design 1.

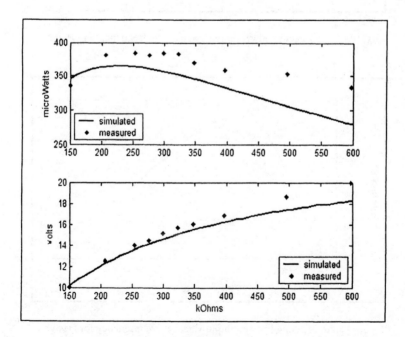

*Figure –5.8.* Measured and simulated output power and voltage versus load resistance for Design 2.

As discussed previously, it is essential for maximal power output that the natural frequency of the generator match the frequency of the input vibrations. Figure 5.9 shows the measured power output versus drive frequency for Design 1. Figure 5.10 shows the measured power output versus drive frequency for Design 2. It appears from the graphs that Design 2 is more sensitive to variations in the drive frequency than Design 1. The most reasonable explanation for this is that the overall damping for Design 2 is lower, and therefore the quality factor is higher.

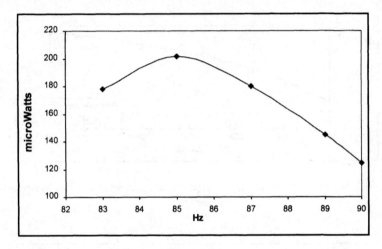

*Figure –5.9.* Measured power output versus drive frequency for Design 1.

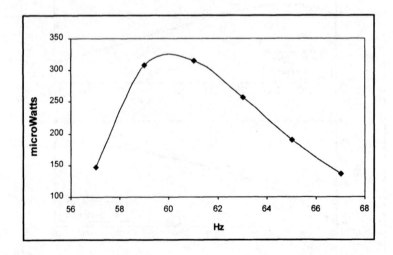

*Figure –5.10.* Measured power output versus drive frequency for Design 2.

# 3.    DISCUSSION OF RESISTIVE LOAD TESTS

The maximum power output and peak voltage values match the simulations rather well. The optimized designs are driving the piezoelectric material harder (closer to its fracture strain). Furthermore, the designs are smaller, and therefore the unaccounted for effect of the clamp may be more significant. Given these two considerations, one may expect that the simulations would not match experiments as well. While there is a greater discrepancy than previously observed, the experimental data still fits quite well. The maximum power output observed from Design 2 was 335 μW compared with a maximum simulated value of 365 μW. The maximum power output observed from Design 1 was 207 μW compared to a maximum simulated value of 200 μW. Furthermore, the data points were tightly grouped around a fitted line that shows the same trend as the simulations.

Note that the maximum simulated power outputs in Figures 7 and 8 do not exactly match those in Tables 4 and 5 of chapter 4. The maximum power output shown in Figure 5.7 is 200 μW compared to 215 μW for Table 4.4. The maximum power in Figure 5.8 is 365 μW compared to 380 μW for Table 4.5. The primary reason is that a damping ratio of 0.025 was used for the simulations shown in Figures 7 and 8 compared with a ratio of 0.02 used for the optimization routine. Also, the measured natural frequencies are lower than the designed natural frequencies. The measured natural frequencies were used in the simulations shown Figures 7 and 8. The higher damping ratios will tend to decrease the power output, and the lower frequencies will tend to increase the power output. The net effect was that the simulated values shown in Figures 7 and 8 are a little lower than those shown in Tables 4 and 5 of chapter 4.

There was a very large mismatch between the designed and measured natural frequency. It should first be mentioned that the parts were actually designed for 100 Hz rather than 120 Hz. A constraint was placed on the natural frequency as part of the optimization routine. However, in order that the optimization would converge more quickly and reliably, a range of 100 Hz to 130 Hz was allowed. In both cases (Design 1 and Design 2) the dimensions generated by the optimization routine resulted in natural frequencies of 100 Hz, as would be expected. After construction, the beam length of Design 2 was measured as 11.3 mm compared to the designed value of 10.7 mm. The measured beam length of Design 1 was 6.5 mm as designed. Finally, the clamp is assumed to be perfectly rigid which is a poor assumption as will be discussed in more detail in the following chapter. The result is that the natural frequency was much lower than the designed value for both designs, and more particularly for Design 2.

A value of 0.18 was used for the coupling coefficient ($k_{31}$) in both the optimizations shown in Tables 4 and 5 of the previous chapter and the simulations shown in Figures 7 and 8. The published value of $k_{31}$ for PZT-5H is 0.44. The coupling coefficient was measured for a prototype made of PZT-5A as described in section 5 of chapter 4. The published $k_{31}$ for PZT-5A is 0.32, and the measured effective value for the bender was 0.12, or 0.375 times the published value. Taking this same ratio, and applying it to the PZT-5H benders, results in an effective coupling coefficient of 0.165. It was found, however, that a slightly higher value of 0.18 results in better matching between calculated and measured output. This coincides with a comment from the manufacturer that the benders with the brass center shim should have slightly better coupling than those with a steel center shim.

There are many similarities between the model developed in chapter 4 for the piezoelectric generator and the generic power conversion model developed in chapter 2, particularly when a simple resistive load is used. If the measured mass, natural frequency, and damping ratio are used as input to the generic model, a quick comparison can be made. Using the values measured from Design 1, the generic model predicts a maximum power output of 239 μW compared to a measured value of 197 μW and a simulated value of 195 μW. Using the values from Design 2, the generic model predicts a maximum power output of 394 μW compared to a measured value of 335 μW and a simulated value of 365 μW.

## 4.       CAPACITIVE LOAD TESTS

The same two prototypes were connected to a capacitive load circuit and driven on the vibrometer. The voltage across the load capacitor was measured and compared to simulated values. The test circuit used is shown in Figure 5.11. The generators were driven at their natural frequency with an acceleration magnitude of 2.25 m/s$^2$ as before. Figure 5.12 shows the measured and simulated voltage across a 1.6 μF load capacitor and the simulated and measured power transfer to the load capacitor versus time from the testing of Design 1. Figure 5.13 shows the same plots with a 3.3 μF load capacitor. Figures 14 and 15 show the measured and simulated voltage and power transfer for Design 2 using 3.3 and 5.5 μF load capacitors. In actual operation a storage capacitor much larger than 1.6 to 5.5 μF would be used. However, smaller capacitors were used for these tests because of the very long simulation times required if much larger capacitors (i.e. super capacitors) are used. As explained in chapter 4, the size of the storage capacitor does not have an effect on the level of power transfer as long as the storage capacitance is two to three orders of magnitude greater than the capacitance of the device. The measured capacitance of each of the two devices tested was about 9 nF, or about 200 to 700 times smaller than the storage capacitors used.

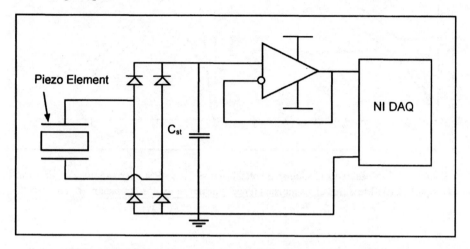

*Figure –5.11.* Measurement circuit for capacitive load tests.

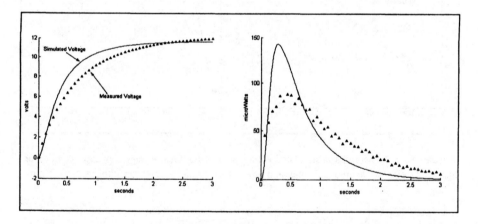

*Figure –5.12.* Measured and simulated voltage across a 1.6 µF storage capacitor vs. time for Design 1 (left). Measured and simulated power transfer to the load capacitor vs. time (right).

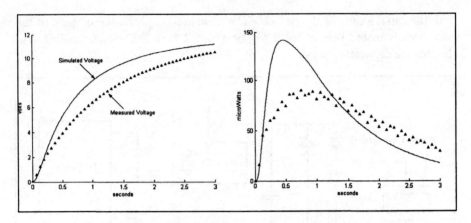

*Figure –5.13.* Measured and simulated voltage across a 3.3 µF storage capacitor vs. time for Design 1 (left). Measured and simulated power transfer to the load capacitor vs. time (right).

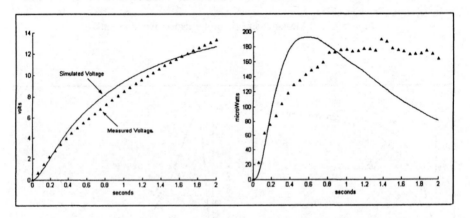

*Figure –5.14.* Measured and simulated voltage across a 3.3 mF storage capacitor vs. time for Design 2 (left). Measured and simulated power transfer to the load capacitor vs. time (right).

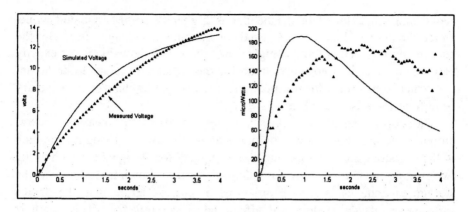

*Figure –5.15.* Measured and simulated voltage across a 5.5 mF storage capacitor vs. time for Design 2 (left). Measured and simulated power transfer to the load capacitor vs. time (right).

## 5.    DISCUSSION OF CAPACITIVE LOAD TEST

There is good agreement between the simulated and measured voltage versus time curves with the exception that in the voltage range of 5 to 10 volts the two curves temporarily deviate from one another. Incidentally, the 5 to 10 volt range corresponds to about 0.5 to 0.8 times the open circuit voltage of the piezo generator. The power transferred to the storage capacitor is a function of $V*dV/dt$ where $V$ is the voltage across the storage capacitor. Therefore, the differences in magnitude and slope between the voltage curves become magnified on the power versus time curves. It will be remembered that the simulations assumed ideal diodes. In reality the voltage drop across the diodes depends on the amount of current flowing through the diodes. As the voltage across the storage capacitor approaches the open circuit voltage, this current is very small. However, at lower voltages, the current is larger, resulting in a larger voltage drop. It is the author's opinion that the unmodeled effect of the diodes is primarily responsible for the discrepancies between the simulated and measured voltage and power versus time curves. While the ideal diode assumption undoubtedly causes some error, it does not change the essential design criteria or the validity of the essential parts of the model.

It should also be noted that the maximum measured voltage for Design 2 actually exceeds the simulated values. The result is that the maximum measured power is still very slightly lower than the maximum simulated power, and the measured power transfer reaches its maximum value later time than the simulated power. The best explanation of this result seems to be that, as explained, the diodes account for the lower slope. However, the actual strain developed in the bender is higher than the simulated value, which would result in a higher final, or maximum, voltage. Just why the maximum measured voltage is higher than the simulated value is not exactly known, however it is not uncommon for the measured voltage to be higher than calculated values for piezoelectric sensors due to changes in boundary conditions (Moulson and Herbert 1997).

The storage capacitance does not seem to affect the measured power output. The maximum power output is about 90 μW for Design 1 for each of the two storage capacitors, and about 180 μW for Design 2 for each of the two storage capacitors. Previous calculations have shown that as long as the storage capacitance is 2 to 3 orders of magnitude larger than the device capacitance, its value does not affect the power transfer. This result is verified by the experimental measurements.

The measured power output is significantly lower for the capacitive load than for the resistive load. The first reason for this is that the devices built and tested were optimized for a resistive load, not for a capacitive load. However, as can be seen from the optimizations performed in chapter 4, the best achievable power transfer is still lower for a capacitive load than for a resistive load even if the design is optimized for a capacitive load.

## 6.       RESULTS FROM TESTING WITH A CUSTOM DESIGNED RF TRANSCEIVER

The generator labeled as Design 2 in Figure 5.2 was used to power a custom designed, low power transceiver. A schematic of the power circuit used in conjunction with the generator is shown in Figure 5.16. Actual part numbers used are labeled on the schematic where appropriate. The physical implementation of the circuit is shown in Figure 5.17. The piezoelectric converter was attached to the vibrometer shown in Figure 5.3 and driven by vibrations at 60 Hz of 2.25 m/s$^2$.

The low power transceiver was designed by Otis and Rabaey (Otis and Rabaey 2002). A block diagram of the receiver and transmitter is shown in Figure 5.18 and the physical transceiver is shown in Figure 5.19. A close-up of one of the custom designed IC's is also shown in the figure. The radio transmits at 1.9 GHz and consumes 10 mA at 1.2 volts. In the test results shown in this section, the transmitter was turned on and broadcast a pure tone. No meaningful information was transmitted. The purpose was to verify the proper functionality of the generator, power circuit, and transmitter, and therefore a simple pure tone was sufficient for the test.

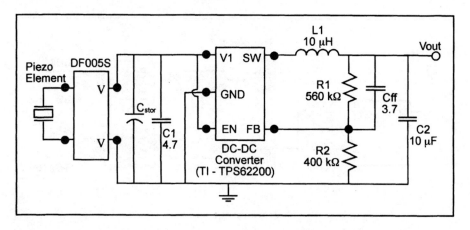

*Figure –5.16.* Schematic of power circuit. V1 is the input voltage to the DC-DC converter, SW is the switching or output pin, GND is the ground pin, EN is the enable pin, and FB is the feedback pin that sets the output voltage.

*Figure –5.17.* Implementation of power circuit.

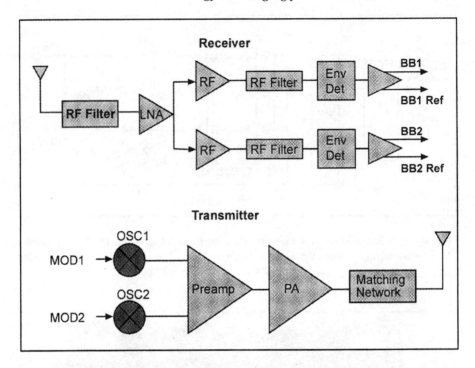

*Figure –5.18.* Block diagram of receiver and transmitter designed by Otis and Rabaey.

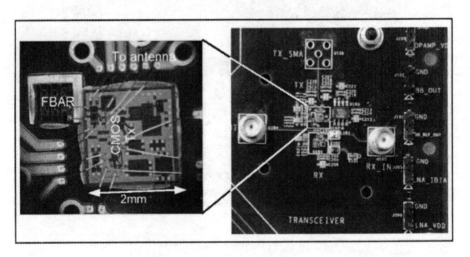

*Figure –5.19.* Custom radio transceiver designed by Otis and Rabaey with close-up of one of the custom chips.

As described earlier, the proposed method of operation of a wireless sensor node is that the radio transmitter and receiver will operate in bursts, being turned off most of the time. The projected duty cycle, the "on" time divided by the total time or $DC = t_{on}/t_{total}$, is typically less than 1% (Rabaey *et al* 2000). During the "off" time, the input capacitor to the DC-DC regulator charges up. During the "on" time, this capacitor is discharged as the power dissipation exceeds the input power. The supportable duty cycle is then given, more or less, by the ratio of input power from the generator to power dissipation when the radio is on. Figure 5.20 shows the results of a test done with a 200 μF input capacitor. In reality, a larger super capacitor on the order of 1 F would probably be used, however the 200 μF capacitor was convenient for the test because it is easier to see the charge / discharge cycle on the input capacitor.

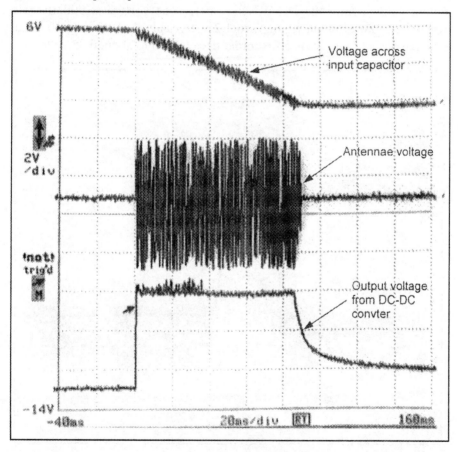

*Figure –5.20.* Test results showing voltage across 200 μF input capacitor, antennae signal, and output of regulator.

The top trace in Figure 5.20 shows the voltage across the input capacitor versus time. Data was acquired for 200 mSec. Although the short time scale makes it difficult to see, the voltage is ramping up at the beginning and end of the trace. The section where the voltage is ramping down is obviously the portion of time for which the radio was on. In this particular case, the switch was closed connecting the input capacitor to the rest of the system and left closed until the voltage across the capacitor fell so far that the DC-DC converter was no longer able to regulate its output. The second trace shows the voltage signal sent out the antennae. In this case a simple pure tone is being transmitted. The transmission frequency is 1.9 GHz. Because of the long time scale (relative to a 1.9 GHz signal) shown, the details of the transmission output cannot be seen. Figure 5.21 shows the frequency spectrum of the transmission signal. Finally the bottom trace shows the output voltage from the DC-DC converter. The voltage is initially zero, when the radio turns on it jumps to 1.2 volts, and when the input falls too low, the output falls off on a first order decay down to about 0.25 volts.

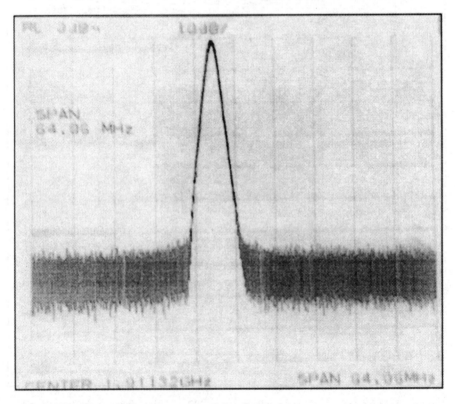

*Figure –5.21.* Frequency spectrum of transmission signal (centered at 1.9 GHz).

Because of the short time scale (relative to the charging time of the input capacitor) of Figure 5.20 it is difficult to tell how fast the input capacitor is being charged. Figure 5.22 shows the voltage across the 200 μF input capacitor. The load is turned off and the input vibrations are the same as for the test shown in Figure 5.20, namely 2.25 m/s² at 60 Hz. Note that it takes about 8.5 seconds to charge from 2 volts to 6 volts. Figure 5.20 showed that it takes about 85 milliseconds for the radio to discharge the input capacitor from 6 volts back down to about 2 volts. Therefore, the supportable duty cycle using this particular generator, vibration source, and radio is 8.5 seconds divided by 0.085 seconds, or 1%.

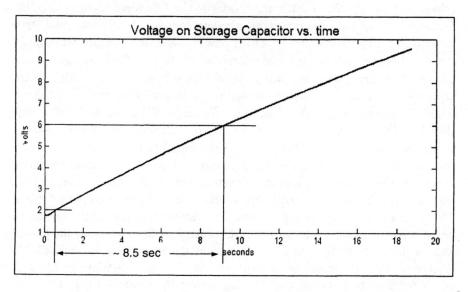

*Figure –5.22*. Voltage across 200 μF input capacitor as it charges up from a vibration source of 2.25 m/s² at 60 Hz.

## 7.  DISCUSSION OF RESULTS FROM CUSTOM RF TRANSCEIVER TEST

The tests to power Otis and Rabaey's transceiver were very successful in that enough power was delivered to be able to support a duty cycle of about 1%, which is the upper end of the projected duty cycle of the entire system. Also, the power delivered was of high enough quality for the transmitter to produce a good, clean signal out. That being said, it should be remembered that while the radio is the highest power portion of an entire sensor node, it does not account for all of the power usage.

The efficiency of the power circuit may be roughly estimated as follows. The power delivered to the radio was 12 mW (10 mA at 1.2 volts) when on, and zero when off. Therefore, the average power delivered to the load is 120 µW. The quiescent current of the DC-DC converter used is about 20 µA. The average input voltage to the DC-DC converter is about 4 volts in this case. So, the quiescent current of the DC-DC converter dissipates about 80 µW of power when on. However, note that the DC-DC converter is off when the radio is off, and the quiescent current in the "off" state is only 1 µA. So, the average power dissipation due to quiescent current is 4.76 µW or roughly 5 µW. Additionally, according to the data sheet, the DC-DC converter is about 90% efficient when on, resulting in an additional average power loss of 12 µW. Finally the diodes in the full wave rectifier account for about a 0.6 volt drop. At an average input voltage of 4 volts, this represents a 13% loss in power. Other power losses, such as current through the feedback resistors, are considered negligible. The average power delivered to the input capacitor is then $120 + 5 + 12 = 137$ µW. This represents 87% of the power produced (13% is lost in the rectifier diodes), which then must be 157 µW. The total efficiency of the power circuit would then be 76%, which isn't bad. However, the full wave rectifier represents more than half the power lost. This could possibly be improved. A little more active circuitry could attempt to ensure that the voltage across the input capacitor remains more or less at the optimal voltage for power transfer from the piezo generator. This could improve the effective produced power above 157 µW. For this particular generator, the best voltage for power transfer is about 8 volts, which would have the additional benefit of reducing the power loss in the rectifier by one half. However, the power lost in the DC-DC converter due to quiescent current would double. Finally, a DC-DC converter designed specifically for this power train would likely be more efficient overall than the off-the-shelf chip used for this test.

One observation is that the input voltage to the DC-DC converter (across the input capacitor) is very noisy. This is due to the high frequency switching of the DC-DC converter and the relatively high input impedance. In actual operation a super capacitor of approximately 1 Farad would likely be used which would decrease the input impedance and reduce this noise. The output voltage from the DC-DC converter is also quite noisy. This could partially be due to the high input impedance. However, the transceiver still operated well with the relatively noisy power signal. Again, a power circuit designed for higher input impedance may also help reduce this noise.

## 8.     RESULTS FROM TEST OF COMPLETE WIRELESS SENSOR NODE

The previous two sections have described the operation of a custom designed RF transceiver. However, the transceiver was not incorporated into a complete functioning wireless sensor node. The power dissipation of currently available wireless sensor nodes is too high to be powered by a vibration to electricity converter of size 1 cm$^3$ or less under vibrations of about 2.25 m/s$^2$. It is likely that the average power consumption of general purpose wireless sensor nodes will soon fall to levels at which they can be powered by a 1cm$^3$ converter from the baseline vibration source used in this study. At present, however, the power consumption is still a factor of 5 to 10 too high. Nevertheless, it is desirable to build and test a complete system to demonstrate feasibility.

Two options are possible. The generators already built could be driven with vibrations of higher amplitude, or a larger generator could be built and driven with low-level vibrations. The latter option was pursued because larger amplitude vibrations run the risk of exceeding the fracture strain of the devices already designed and built. An additional consideration is that driving the generator with vibrations of higher amplitude will increase the open circuit voltage produced by the generator, which means that the best operating voltage at the input to the DC-DC converter will be higher. Using the generator labeled as Design 2, the resulting input voltage would be about 12 volts. Such a large difference between the input and output voltage greatly limits the number of commercial DC-DC converters that can be used.

A larger generator was designed and built. This generator is shown in Figure 5.23 and will be referred to as Design 3. The size of the generator is about 3 cm by 2 cm by 0.8 cm, or 6 times larger than Design 2. The proof mass is 52.2 grams or 6.4 times larger than Design 2. The generator was used to power the small general purpose, programmable wireless sensor node shown in Figure 5.24 (Warneke *et al* 2001)[1].

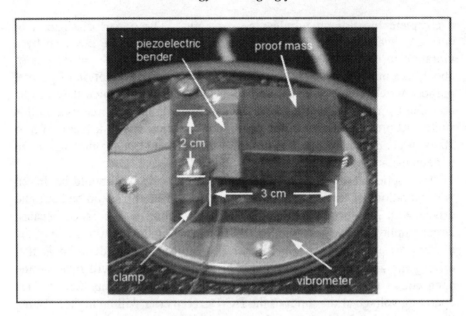

*Figure −5.23*. Larger generator built to power a complete wireless sensor node.

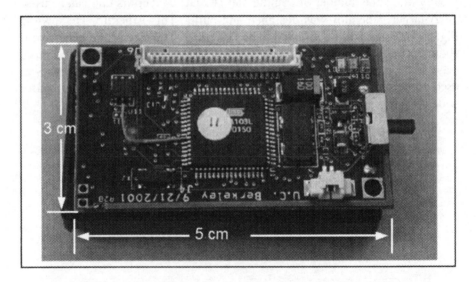

*Figure −5.24*. Complete programmable wireless sensor node.

The generator was designed to have roughly the same open circuit voltage as Design 2, but to output much more current. It was to be designed to run at 120 Hz, however, due to two factors, the actual resonant frequency is only 40 Hz. The first factor is that the clamp is more compliant than accounted for. The second factor is that the material of the wrong thickness was used in constructing the device resulting in lower stiffness and higher capacitance. The device capacitance was measured as 171 nF. The device was driven with accelerations of 2.25 m/s$^2$ (as in previous tests) at 40 Hz. The power output versus load resistance is shown in Figure 5.25 for the case when the generator is terminated with a simple resistor. Note that the maximum power is 1700 μW and occurs at a load resistance of 18 kΩ.

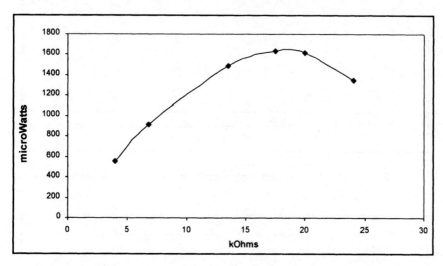

*Figure –5.25.* Power vs. load resistance for the generator labeled Design 3.

The same power circuit shown in Figure 5.17 was used to connect the generator to the wireless sensor node with two small alterations. The feedback resistors for the DC-DC converter were altered to output the 3 volts needed by the sensor node rather than the 1.2 volts used in the previous test. Secondly, an input capacitor of 10 mF was used rather than 200 μF because of the greater power production and dissipation. Again, in real operation a super capacitor of about 1 F would be used, but this size capacitor makes it difficult to see the voltage variations that correlate to system operation. Figure 5.26 shows the charge-up of the 10 mF input capacitor without the load connected. Figure 5.27 shows the power transfer from the converter to the input capacitor versus time. As can be seen, when the voltage across the capacitor reaches an appropriate voltage (about 0.5 times the open circuit voltage) the power transfer is about 700 μW.

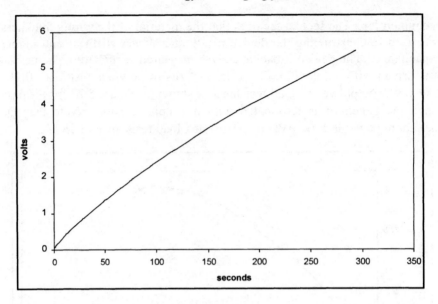

*Figure –5.26.* Voltage across the 10 mF input capacitor versus time with load disconnected.

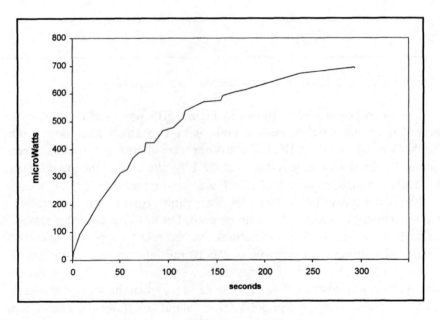

*Figure –5.27.* Power transfer to the 10 mF input capacitor versus time with load disconnected.

A program that measures one of the sensor inputs 10 times per second and transmits the reading was written and downloaded to the wireless sensor node shown in Figure 5.24. Each transmitted packet contains only one sensor reading. The node accepts analog sensor inputs from 0 to 3 volts. The voltage across the input capacitor was divided by 4 with a resistive divider and wired to the sensor input. The voltage across the input capacitor was also directly measured with a data acquisition system for comparison. Because the power consumption of the node is far greater than 700 µW, a switch was placed in between the input capacitor and DC-DC converter in order to turn the system on and off in the same manner done previously. Figure 5.28 shows both the directly measured voltage across the input capacitor and the transmitted voltage multiplied by 4. Figure 5.29 shows the calculated power consumption of the wireless sensor node. The power consumption was calculated from the voltage profile across the input capacitor. The power out of the capacitor is simply $P_{out} = V_{in}*I_{out} = -C_{in} V_{in} dV_{in}/dt$. Subtracting the power dissipation of the DC-DC converter, the power lost through resistive dividers, and the power transferred to into the input capacitor from the generator, the result is the power dissipation of the wireless sensor node as shown in Figure 5.29.

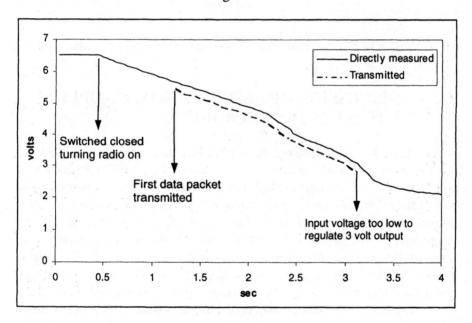

*Figure –5.28.* Directly measured and transmitted input voltage versus time.

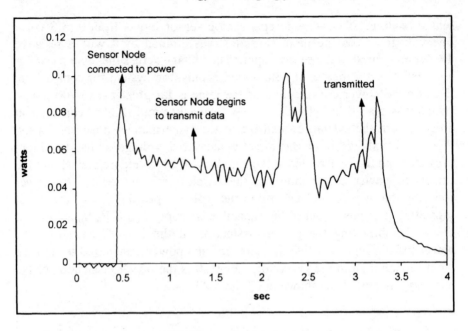

*Figure –5.29.* Power dissipation versus time for the wireless sensor node.

## 9.    DISCUSSION OF RESULTS FROM COMPLETE WIRELESS SENSOR NODE

The generator was designed for a 0.02 inch thick bender. The bender actually used was 0.015 inches thick. Because the resonant frequency is proportional to the thickness cubed, the resulting resonant frequency should be $(.015/.02)^3$ or 0.42 times the designed frequency. This would result in a resonant frequency of 50 Hz. The remaining discrepancy is due to compliance of the clamp. While the input vibrations are not exactly the same as those used for Design 2, they are close enough to provide a reasonably good comparison. Remembering that Design 3 (the larger design) is about 6 times larger than Design 2 (the mass is actually 6.4 times that of Design 2), the power output of Design 3 should be about 6 times greater. The maximum power through a resistive load for Design 3 is 1700 µW compared to 335 µW for Design 2 resulting in a power output ratio of 5:1. Further investigation would be necessary to determine why the power ratio does not track exactly with the mass ratio. However, a 20% percent discrepancy is not unexpected and does not significantly contradict the fact that power output is proportional to mass.

The voltage profiles shown in Figure 5.28 indicate an offset of about 0.25 volts between the directly measured voltage and the transmitted voltage. It is possible that the voltage divider across the input capacitor was not exactly a ¼ divider due to parallel resistances. It is also possible that the voltage results from the acquisition of the signal by the wireless sensor node. In either case, the reason is not particularly important for this test. The purpose was to demonstrate a fully functional system, which the test does effectively. A second observation is that the wireless sensor node takes about 0.75 seconds to start up and begin acquiring and transmitting data. In actual operation a hard switch would not be used to shut the node down. The node could be programmed to go into a very low power sleep mode and turn on at some predetermined duty cycle. However, the implementation of such a program is beyond the scope of this test.

An approximate supportable duty cycle can be calculated based on the data shown in Figures 26 and 28. Figure 5.26 shows that the generator charges the input capacitor from 3 volts to 5 volts in 130 seconds. Figure 5.28 demonstrates that it takes the sensor node about 1.3 seconds to dissipate through this same range (5 volts back down to 3 volts). Therefore, the supportable duty cycle is about 1% for this system. More intelligent power circuitry that would attempt to maintain the input voltage at its optimal value would considerably improve the supportable duty cycle.

In the same manner that was done for the previous test reported, the approximate efficiency of the system can be quickly estimated. In this case the power delivered to the wireless sensor node ranges from 50 mW to 100 mW with an average value of about 60 mW. At a duty cycle of 1% the average power consumption of the node would be 600 μW (assuming that power dissipation is zero when "off"). Just as described previously in section 7 the power dissipation in the DC-DC converter due to quiescent current is about 5 μW. The converter is roughly 90% efficient when on, resulting in an additional average power loss of 60 μW. Finally the diodes account for account for about a 10.7% loss in power due to voltage drop. (The average voltage across the input capacitor is about 5 volts for this example.) Current through feedback resistors is considered negligible. The average power delivered to the input capacitor is then 600 + 5 + 60 = 665 μW. This represents 89.3% of the power produced, which then must be 745 μW. The total efficiency of the power circuit would then be 80.6%. Again, over half of the power loss is due to the rectification diodes. The efficiency of the circuit could be improved in exactly the same manner as described in section 7.

## 10.    CONCLUSIONS

Four basic sets of tests have been performed with three different piezoelectric vibration-to-electricity converters. Two 1cm$^3$ converters were tested using a simple resistive load to characterize the maximum power generation. Secondly, the two converters were tested with a purely capacitive load to characterize the power transfer in a more realistic situation. One of the converters was then used to power a custom design radio transceiver that consumes 12 mW when on. Finally, a third, larger converter was built and used to power a complete wireless sensor node that draws approximately 60 mW when on. The conclusions from these tests are summarized below.

1. The maximum demonstrated power density from a vibration source of 2.25 m/s$^2$ at 60 Hz is 335 µW/cm$^3$.
2. The maximum measured power transfer to a capacitive load is 180 µW/cm$^3$, or a little over half the power dissipated by a purely resistive load.
3. A 1 cm$^3$ generator has been successfully used to power a custom design RF transceiver. The generator can sustain about a 1% duty cycle for this transceiver.
4. A larger, 6 cm$^3$, generator has been built and used to test a complete programmable wireless sensor node. The sustainable duty cycle was 1%.
5. The efficiency of the power circuitry used is approximately 75% - 80%. Significant improvement can be made on the design of the power electronics.

---

[1] The wireless sensor node shown in Figure 5.24 was developed by a group of researchers at UC Berkeley. It is generally referred to as the "Mica Mote". The same group has since developed another, smaller wireless sensor node referred to as a "Dot Mote", which has about the same footprint of the US quarter with a thickness of about 1cm. Both the "Mica Mote" and "Dot Mote" are now manufactured by Crossbow Technology, Inc. Additionally, very similar wireless nodes are now manufactured and sold by Dust Inc.

# Chapter 6

# ELECTROSTATIC CONVERTER DESIGN

Chapters 4 and 5 have dealt with piezoelectric converters. The direction will now change to the consideration of electrostatic converters. This chapter will deal with the modeling, design, and optimization of electrostatic converters. The following chapter will discuss the processing and fabrication of electrostatic converters. And finally, chapter 8 will cover test results.

## 1.     EXPLANATION OF CONCEPT AND PRINCIPLE OF OPERATION

Recalling the discussion of electrostatic conversion presented in chapter 3, section 3, the variable capacitor is the basis of power conversion. Mechanical vibrations driving a capacitive structure cause the capacitance, and thus the energy stored in the capacitor, to change.

Reference was made to both charge constrained and voltage constrained conversion. Again, Meninger et al (Meninger et al 2001) give a good explanation of the merits of both charge and voltage constrained conversion. In theory, slightly more power could be produced from a voltage constrained system. This conclusion however assumes that the change in capacitance of the system is limited by a maximum allowable voltage rather than by the kinetic energy imparted by the driving vibrations. This assumption will often not be valid, and so the only advantage of a voltage constrained system is often not operative. The primary disadvantage of the voltage constrained system is that it requires two separate voltage sources for conversion to take place. The charge constrained system is much simpler in that it only requires one separate voltage source. Therefore, a charge constrained system seems the best of the two alternatives, and has been chosen as a target system for this study. Meninger *et al* have also chosen a charge constrained system, however, no mention is made of the underlying assumption inherent in calculating the power advantage of the voltage constrained system.

A simplified circuit for an electrostatic generator using charge constrained conversion is shown in Figure 6.1. This circuit is useful for power output calculations and demonstrates the basic function of energy conversion although it is not entirely realistic. A pre-charged reservoir, which could be a capacitor or rechargeable battery, is represented as the input voltage source $V_{in}$. The variable capacitor $C_v$ is the variable capacitance structure, and $C_{par}$ is the parasitic capacitance associated with the variable capacitance structure and any interconnections. When $C_v$ is at its maximum capacitance state ($C_{max}$), switch 1 (SW1) closes, and charge is transferred from the input to the variable capacitor. The capacitive structure then moves from its maximum capacitance position to the minimum capacitance position ($C_{min}$) with both switches open. The result is that the energy stored on $C_v$ increases. At minimum capacitance, switch 2 (SW2) closes and the charge stored on $C_v$ (now in a higher energy state) is transferred onto the storage capacitor $C_{stor}$. The mechanical vibrations have done work on the variable capacitor causing an increase in the total energy stored in the system.

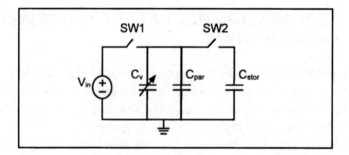

*Figure –6.1.* Simple circuit representation for an electrostatic converter.

In reality, the switches would either be diodes, or transistors with an inductor, and some method of returning a portion of the charge put onto the storage capacitor to the input reservoir would need to be employed. Researchers at MIT (Amirtharajah and Chandrakasan 1998, Amirtharajah 1999, Meninger *et al* 1999, Amirtharajah *et al* 2000, Meninger *et al* 2001) have developed a circuit to accomplish this task. Detailed development of the power circuit is considered outside the scope of the current study. The circuit shown in Figure 6.1 is sufficient to obtain power estimates and to use a basis for the design of the physical variable capacitance structure.

## 2.    ELECTROSTATIC CONVERSION MODEL

The increase in energy stored in the variable capacitor per cycle is given by the equivalent expressions in equations 7.1a and 7.1b.

$$E = \frac{1}{2}V_{in}^2(C_{max} - C_{min})\left(\frac{C_{max} + C_{par}}{C_{min} + C_{par}}\right) \tag{6.1a}$$

$$E = \frac{1}{2}V_{max}V_{in}(C_{max} - C_{min}) \tag{6.2a}$$

where $V_{max}$ represents the maximum allowable voltage across a switch.

Depending on the specific implementation of the switches, the physical design, and the input vibrations, $V_{max}$ may be a very limiting constraint. If it can be determined that the maximum allowable voltage will be a limiting constraint, then equation 6.1b may be the more useful of the two. Otherwise, equation 6.1a will be more useful in designing the system. The power output is, of course, just given by the energy per cycle multiplied by the frequency of operation, which will necessarily be the frequency of the input vibrations.

The energy transfer per cycle is highly dependent on the ratio of maximum to minimum capacitance. It is, therefore, important to note that actual distance of travel of the variable capacitance structure, and therefore the value of $C_{max}$ and $C_{min}$, is determined by both the mechanical dynamics of the system and the design of the capacitive structure. A schematic of a vibrating mechanical system was shown in chapter 2 as Figure 2.4. The basic premise of the system is that there are two dampers, a mechanical damper representing pure loss, and an electrical damper representing the energy removed from the mechanical system and transferred to the electrical system. In chapter 2 it was assumed that both of these dampers were linear and proportional to velocity. Although the same schematic representation will be used to develop the dynamic model for electrostatic converters, it cannot be assumed that either damper is linear or proportional to velocity.

The equation of motion for the mechanical portion of the system, based on the schematic of Figure 2.4, is given here in equation 6.2.

$$m\ddot{z} + f_e() + f_m() + kz = -m\ddot{y} \tag{6.2}$$

where $m$ is the mass of the oscillating capacitive structure, $k$ is the stiffness of the flexures on the capacitive structure, $z$ is the displacement of the capacitive structure, $y$ is the input vibration signal, $f_e(\ )$ represents the electrically induced damping force function, and $f_m(\ )$ represents the mechanical damping force function.

The capacitance of the variable capacitor at a given time $(t)$ is determined by the displacement of the structure $(z)$ and the specifics of the design. The amount of energy per cycle that is removed from the mechanical system, and stored in the electrical system is given by equation 6.3.

$$E = \int_{0}^{2\pi/\omega} f_e(\ )dz \tag{6.3}$$

where $\omega$ is the frequency of oscillation.

The expression for energy per cycle given by equation 6.3 is equivalent to that shown in equation 6.1a. Consider a simple example of a variable capacitor consisting of two square plates. The capacitor changes capacitance as one plate, attached to springs, oscillates between values $z_{min}$ and $z_{max}$ where $z$ is the distance between the two plates. The form of $f_e(\ )$ for this example is given by the following expression.

$$f_e(\ ) = \frac{Q^2}{2\varepsilon_0 A} \tag{6.4}$$

where, $Q$ is the charge on the capacitor, which is constrained to be constant, $\varepsilon_0$ is the dielectric constant of free space, and $A$ is the area of the capacitor plates.

Note that for this example, $f_e(\ )$ is constant and not a function of $z$, however this is not always the case. Substituting equation 6.4 into equation 6.3 and solving yields the following expression.

$$E = \frac{Q^2}{2\varepsilon_0 A}\left(z_{max} - z_{min}\right) \tag{6.5}$$

In deriving equation 6.5 it was assumed that all the charge is removed from the variable capacitor as it returns from the $C_{min}$ (or $z_{max}$) state back to the $C_{max}$ (or $z_{min}$) state. Noting that $C_{max} = \varepsilon_0 A/z_{min}$, $C_{min} = \varepsilon_0 A/z_{max}$, and $Q = C_{max}*V_{in}$, equation 6.5 can be easily reduced to the following form.

$$E = \frac{1}{2}V_{in}^2 \left(C_{max} - C_{min}\right)\left(\frac{C_{max}}{C_{min}}\right) \tag{6.6}$$

Equation 6.6 is the same expression as equation 6.1a neglecting the parasitic capacitance. Note also that $(z_{max} - z_{min})$ is nothing more than the AC magnitude of $z$ (the distance between plates). If it is assumed that the mechanical damping is linear viscous damping ($f_m( ) = b_m\dot{z}$), then the AC magnitude of $z$ is given by equation 6.7.

$$|Z| = \frac{m\omega}{b_m}|Y| \tag{6.7}$$

where $|Y|$ is the displacement magnitude of the input vibrations, $b_m$ is a constant damping coefficient, and $\omega$ is assumed to be equal to the natural frequency of the capacitive structure.

Substituting equation 6.7 into equation 6.5 and replacing displacement magnitude with acceleration magnitude ($|A_{in}| = \omega^2|Y|$) yields the expression in equation 6.8.

$$E = \frac{mQ^2}{2\omega b_m \varepsilon_0 A}|A_{in}| \tag{6.8}$$

where $A_{in}$ is the acceleration magnitude of the input vibrations, $A_{in} = Y/\omega^2$.

Equation 6.8 clearly shows that the energy converted per cycle is linearly proportional to the mass of the system. This same conclusion was obtained in the development of the generic energy conversion model presented in chapter 2. Therefore, maximizing the mass of the system becomes an important design consideration. Furthermore, energy output is inversely proportional to frequency assuming that the acceleration magnitude of the input vibrations does not increase with frequency. Again, this conclusion is consistent with the generic model developed in chapter 2. However, in contrast to the generic model, equation 6.8 would imply that power output is proportional to $A_{in}$ rather than to $A_{in}^2$.

The parasitic capacitance has two effects on the system. First, the ratio of maximum voltage ($V_{max}$) across $C_v$ to the input ($V_{in}$) voltage, given by equation 6.9, is affected by $C_{par}$.

$$\frac{V_{max}}{V_{in}} = \frac{C_{max} + C_{par}}{C_{min} + C_{par}} \tag{6.9}$$

This can be important in two ways. If the ratio of maximum to minimum voltage is too small, the system will not function. For example if the two switches are implemented as diodes, the $V_{max} - V_{min}$ must be at least large enough to overcome the forward voltage across the diode. In this case $C_{par}$ would need to be minimized in order to maximize the $V_{max}/V_{min}$ ratio. In the unlikely case that $V_{max}$ is too large (i.e. greater than the maximum allowable voltage for the system), a larger $C_{par}$ would reduce $V_{max}$, and therefore be desirable. Second, a larger $C_{par}$ will result in a larger electrostatic force on the oscillating mass. In other words, it will result in more electrically induced damping. (This will be demonstrated more fully when specific designs are considered.) Meninger *et al* (Meninger *et al* 2001) make the assertion that a large $C_{par}$ will improve the energy conversion per cycle. This is only true if the dynamics are such that enough displacement can still be achieved with the larger $C_{par}$ to reach the maximum allowable voltage. This is unlikely however, because a large $C_{par}$ will increase the overall damping of the system, thus reducing the displacement of the variable capacitor. In general, it is usually best to try to reduce the parasitic capacitance, and then set the desired level of electrically induced damping with either the input voltage ($V_{in}$) or mechanical design parameters.

Specific power output estimates and dynamic simulations cannot be performed without first specifying the specific design concept. The choice of a design concept will determine the forms of $f_e(\ )$ and $f_m(\ )$, which will complete the model, presented and allow specific calculations and optimization based on the model.

## 3.      EXPLORATION OF DESIGN CONCEPTS AND DEVICE SPECIFIC MODELS

As mentioned in chapter 3, the primary reason to pursue electrostatic energy conversion is the ease with which electrostatic converters can be implemented with silicon micromachining technology (or MEMS). A MEMS implementation has a few advantages. First, it has the potential to be tightly integrated with silicon based microelectronics. Second, equipment and processes to mass produce silicon micromachined devices are readily available. And third, if it ever becomes attractive to drastically reduce the size of converters, further miniaturization is readily accomplished with MEMS technology. Because the potential for MEMS implementation is the only advantage of an electrostatic converter over a piezoelectric converter, it only makes sense to consider designs that can readily be manufactured in micromachining processes.

Micromachined devices are generally planar devices (2½D) fabricated on the surface of a silicon substrate (surface micromachining). The silicon substrate can also be etched to create devices (bulk micromachining). New innovative micromachining techniques are continually being developed. A detailed discussion of micromachining is beyond the scope of this thesis. The reader is referred to Madou (Madou 1997) for a full treatment of micromachining technology applied to MEMS. For the purposes of this thesis, it will be assumed that the reader is somewhat familiar with basic micromachining processes. As demonstrated above, the maximum capacitance of the device is a key parameter to effective power conversion. It is therefore desirable to target a process that can produce devices with large capacitances. A very thick device layer, and a high aspect ratio are therefore necessary. A process in which the devices are fabricated in the top layer of a Silicon-On-Insulator (SOI) wafer is capable of producing very thick structures (up to 500 μm or more). Furthermore, the Deep Reactive Ion Etching (DRIE) process used to create the devices is capable of producing features with aspect ratios up to 50:1. For these reasons, a SOI process will be targeted for the design of MEMS based electrostatic converters. The specific details of the fabrication process will be covered in the following section.

Three topologies for micromachined variable capacitors are shown in Figure 6.2. The dark areas are fixed by anchors to the substrate, while the light areas are released structures that are free to move. The first device at the top will be referred to as an in-plane overlap converter because the change in capacitance arises from the changing overlap area of the many interdigitated fingers. As the center plate moves in the direction shown, the overlap area, and thus the capacitance, of the fingers changes. The second device will be referred to as an in-plane gap closing converter because the capacitance changes due to the changing dielectric gap between the fingers. The third device shown will be referred to as an out-of-plane gap closing converter. Note that the figure shows a top view of the first two devices, and a side view of the third device. This third device oscillates out of the plane of the wafer, and changes its capacitance by changing the dielectric distance between two large plates. A few representative dimensions are shown in the figure.

*Energy Scavenging for Wireless Sensor Networks*

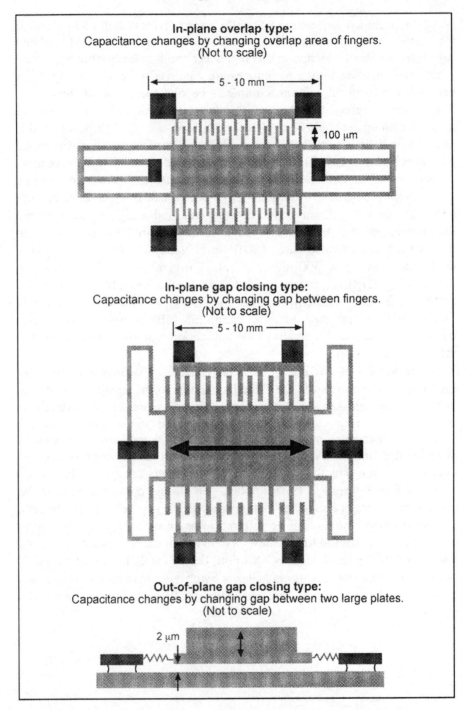

*Figure –6.2.* Three possible topologies for micromachined electrostatic converters.

## 3.1      Out-of-plan gap closing converter

The out-of-plane gap closing type converter will be considered first. The exact expression for the mechanical damping term is give by equation 6.10.

$$f_m() = \frac{16\mu W^3 L}{z^3}\dot{z} \tag{6.10}$$

where $\mu$ is the viscosity of air (the value of $\mu$ is proportional to the pressure), $W$ is the width of the large plate, and $L$ is the length of the plate.

Note that the interpretation of $z$ in this equation is slightly different than as shown in Figure 2.4 in that $z$ represents the separation of the two plates making up the capacitor. Thus $z$ is the sum of the initial separation and the deflection of the flexures. The capacitance of this structure is given by the following expression.

$$C_v = \frac{\varepsilon_0 WL}{z} \tag{6.11}$$

where $\varepsilon_0$ is the dielectric constant of free space.

Finally, the expression for the electrostatic force induced is given by equation 6.12.

$$f_e() = \frac{-Q^2}{2\varepsilon_0 WL} \tag{6.12}$$

where $Q$ is the charge on the variable capacitor.

Because the charge is held constant during the motion of the structure, the electrostatic force is constant. When the switches close, the amount of charge on the capacitor changes, but this happens very fast and can be considered to be instantaneous from the viewpoint of the mechanical subsystem.

Equations 7.10 – 7.12 demonstrate one of the problems with the out-of-plane gap closing converter. In order to obtain a large capacitance change, $z$ must become very small, or the plates must move very close together. However, as the fluid damping force is proportional to $1/z^3$, the loss becomes very large as the plates move close together. This problem may be alleviated somewhat by packaging the device under very low pressure. However, another serious problem exists with this design concept. As the plates get close together, surface interaction forces will tend to make them stick together shorting the circuit and possibly becoming permanently attached. It is very difficult to design mechanical stops to prevent this from happening with the out-of-plane topology.

## 3.2 In-plane gap closing converter

The in-plane gap closing converter considerably improves the latter problem mentioned with the out-of-plane converter. Because the motion is now in the plane of the wafer, mechanical stops can be easily incorporated with standard fabrication processes, and therefore, the minimum dielectric gap, and thus the maximum capacitance can be precisely controlled. The expression for the fluid damping term for the in-plane gap closing type converter is given by equation 6.13.

$$f_m() = \left( \frac{\mu A}{d_0} + 16\mu N_g L_f h^3 \left( \frac{1}{(d-z)^3} + \frac{1}{(d+z)^3} \right) \right) \dot{z} \qquad (6.13)$$

where $A$ is the area of the center plate, $d_0$ is the vertical distance between the center plate and the substrate underneath, $N_g$ is the number of gaps per side formed by the interdigitated fingers, $L_f$ is the length of the fingers, h is the thickness of the device, and $d$ is the initial gap between fingers.

Note that z in this expression is the deflection of the flexures. The capacitance of the structure is given by the following expression:

$$C_v = N_g \varepsilon_0 L_f h \left( \frac{2d}{d^2 - z^2} \right) \qquad (6.14)$$

The expression for the electrostatic force induced is given by equation 6.15.

$$f_e() = \frac{Q^2 z}{2N_g d\varepsilon_0 L_f h} \qquad (6.15)$$

Note that the electrostatic force is proportional to the deflection of the flexure, and thus acts much like a mechanical spring except that the electrostatic force operates in the opposite direction.

While the fluid damping is still quite high for this design, large differences in capacitance can be generated and precise control of the maximum capacitance is possible if mechanical stops are included in the design.

## 3.3 In-plane overlap converter

The expression for fluid damping for the overlap in-plane converter is given by expression below.

$$f_m() = \frac{N_g \mu L_f h}{d} \dot{z}$$ (6.16)

where $d$ is the dielectric gap between fingers.

Equation 6.16 is actually in the standard form for linear viscous damping. The capacitance for the structure is given by equation 6.17.

$$C_v = \frac{N_g \varepsilon_0 L_f (z + z_0)}{d}$$ (6.17)

where $z_0$ is the initial overlap distance of interdigitated fingers.

The expression for the electrostatic force induced is given by the following expression.

$$f_e() = \frac{Q^2 d}{2N_g \varepsilon_0 h (z + z_0)^2}$$ (6.18)

## 4. COMPARISON OF DESIGN CONCEPTS

A useful comparison between in-plane overlap and gap closing converters can be made without performing simulations that take into account the full dynamics of the systems. Estimates of power output per cm$^3$ based only on geometry and the relationship in equation 6.1a are graphed against maximum flexure (spring) deflection in Figures 6.3 – 6.5. The input voltage used for this comparison was 5 volts. Realistic assumptions were made about the minimum finger thickness and gap between fingers based on the standard fabrication technology. The distance between fingers for the in-plane overlap converters was assumed to be 1μm, and the minimum dielectric gap for both types of gap closing converters was assumed to be 0.25 μm. A device thickness of 50 μm was used for all three types of devices which is a realistic thickness based the targeted SOI process that will be described later.

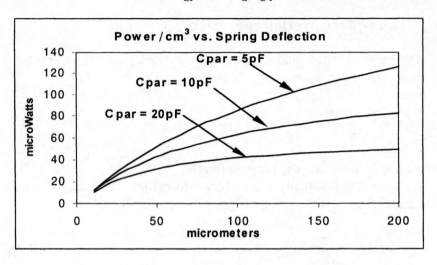

*Figure –6.3.* Power density vs. flexure deflection for out-of-plane gap closing converter with three different parasitic capacitances.

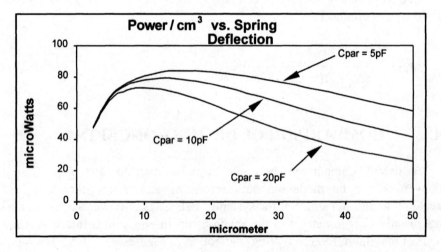

*Figure –6.4.* Power density vs. flexure deflection for in-plane gap closing converter with three different parasitic capacitances.

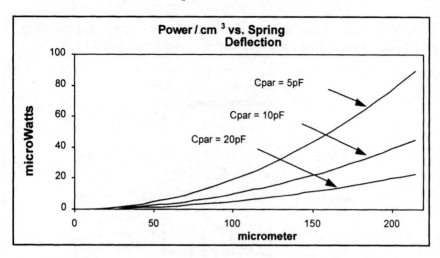

*Figure –6.5.* Power density vs. flexure deflection for in-plane overlap converter with three different parasitic capacitances.

It should be noted that for the in-plane gap closing converter the number of fingers that can be fabricated is a function of the maximum deflection of the flexures because the fingers must be spaced far enough apart to accommodate the displacement. Therefore a higher spring deflection results in a lower maximum (and minimum) capacitance. This is not true for in-plane overlap or out-of-plane gap closing converters. The result is that for both in-plane overlap and out-of-plane gap closing converters a larger spring deflection always results in more power out. However, there is an optimal travel distance for in-plane gap closing converters as can be seen Figure 6.4.

All three types of converters are capable or roughly the same power output. The out-of-plane gap closing converter produces the highest power output especially at low parasitic capacitances. The estimates presented here are not based on optimized designs, but on engineering judgment and the realistic constraints of the microfabrication. However, very useful trends can be seen from these estimates.

First, the maximum power output for in-plane overlap and out-of-plane gap closing converters occurs at very high spring deflections. This issue is even more acute for the overlap converter because of the upward curvature of the power traces compared with the downward curvature for out-of-plane gap closing converters. Such large deflections raise concerns about the stability of the system. In the case of the overlap type converter, if the deflections are very large (on the order of 100 μm) and the gap is very small (on the order of 1 μm), only a small moment induced by out-of-axis vibrations would be necessary to cause the fingers to touch and short the circuit. This potential problem is illustrated in Figure 6.6. The optimal spring deflection for the in-plane gap closing converters is around 10 to 15 μm, which is very realistic and will not pose much of a stability problem.

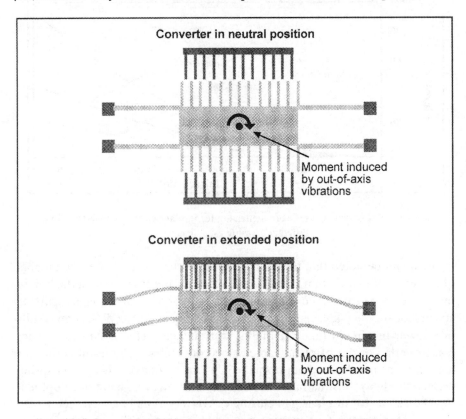

*Figure –6.6.* Illustration of stability problem with in-plane overlap converters.

Second, the out-of-plane gap closing converters are far less sensitive to the parasitic capacitance. A parasitic capacitance of 5 pF is very optimistic, and would likely not be possible. A parasitic on the order of tens of picoFarads is much more likely. Finally, remember that there is no guarantee that the very large spring deflections shown for in-plane overlap and out-of-plane gap closing converters can in fact be obtained. The maximum obtainable deflections depend on the mechanics of the system, which were not included in these estimates. These preliminary estimates, however, point out the two considerations discussed better than a full dynamic simulation.

A full simulation of the three designs of Figures 3 – 5 will provide more details that can be used as a basis for comparison. Simulations were performed in Matlab using equations 7.1,7.2, and 7.10 – 7.18. As was the case with simulations for piezoelectric converters, input vibrations of 2.25

m/s$^2$ at 120 Hz were used for all simulations. Figure 6.7 shows some results of a simulation of the out-of-plane gap closing converter used to generate the power estimates in Figure 6.3. Figure 6.7 shows the voltage on the storage capacitor (100 pF in this case) and the voltage on the variable capacitor, C$_v$, versus time. The traces clearly demonstrate the basic charge pump function of the simple simulation circuit shown in Figure 6.1. This simulation was performed with an input voltage of 5 volts, an ambient pressure of 0.01 atmospheres, and a parasitic capacitance of 20 pF. The switches are assumed to be ideal in that they turn on and off instantaneously and switching loss is neglected.

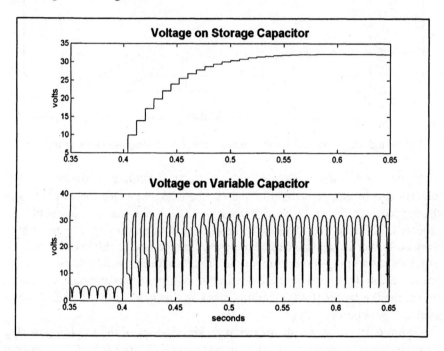

*Figure –6.7.* Voltage on storage capacitor and variable capacitor vs. time, demonstrating the charge pump like function of the converter.

As discussed earlier, the out-of-plane topology suffers from very high squeeze film damping forces. At atmospheric pressure these damping forces dominate the system, and so most of the kinetic energy of the system is lost and very little power output is available. As various methods do exist to package MEMS structures at reduced pressures (Hsu 2000, Chang and Lin 2001), the system was simulated at a variety of different pressures. Figure 6.8 shows the output power per cm$^3$ vs. pressure in atmospheres. At .001 atmospheres (or 0.76 torr), the power output is 20 µW/cm$^3$, which may be in the useful range. At atmospheric pressure the power output is on the order of 1 nW/cm$^3$, which is far too low to be of any use.

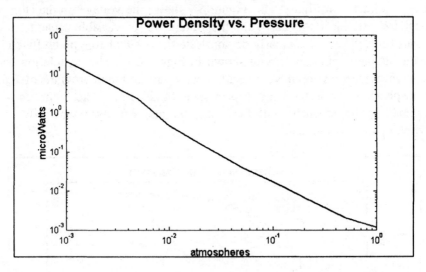

*Figure –6.8.* Power per cm3 vs. air pressure for an out-of-plane gap closing converter.

Figure 6.9 shows power output versus minimum dielectric gap (determined by the placement of the mechanical stops) for the in-plane gap closing converter for three different device thicknesses. As would be expected, higher power occurs for smaller minimum dielectric gaps, which produce higher maximum capacitances. A larger device thickness results in larger fluid damping forces, higher electrostatic forces, and higher maximum capacitances. It appears that the advantage of higher maximum capacitances outweighs the larger fluid damping forces in this case as the thicker devices produce more power output.

Figure 6.10 shows results of dynamic simulations of the in-plane overlap converter used to generate the power estimates in Figure 6.5. Power output versus fabricated distance between fingers (dielectric gap) is shown for three different device thicknesses. This simulation was also performed with an input voltage of 5 volts, an ambient pressure of 0.01 atmospheres, and a parasitic capacitance of 20 pF. As expected, lower dielectric gaps again result in higher power output. In this case the device thickness makes very little difference. The increased damping cancels the benefit of higher maximum capacitances. The most important thing to notice from the figure is that even at the very unrealistic dielectric gap of 0.2 µm the power output is only about 11 µW/cm$^3$, which is two to three times lower than the in-plane gap closing converter. The reason for the lower power output is that the spring deflections necessary to obtain higher output power densities are simply not achievable given the input vibrations and the simulated dynamics of the system.

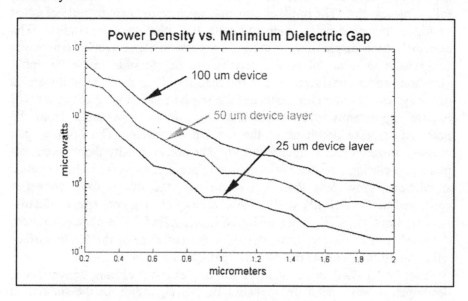

*Figure –6.9.* Power output vs. minimum allowable dielectric gap for an in-plane gap closing converter for three different device thicknesses.

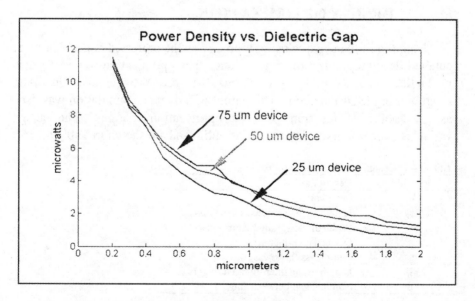

*Figure –6.10.* Power output vs. fabricated dielectric gap for an in-plane overlap converter for three different device thicknesses.

Given the simulations and power estimates presented above the in-plane gap closing topology seems to be the best of the three options for the following reasons: The in-plane gap closing converters are capable of equal or higher power density compared with the other two topologies. The extremely high displacements needed to make in-plane overlap converters comparable in terms of power density are not feasible given the input vibrations under consideration and the realities of the mechanical dynamics of the system. Furthermore, because the in-plane gap closing converter will require significantly smaller displacements, it will not suffer from the potential stability problems of the overlap converter. The in-plane gap closing converter has a significantly higher power density than the out-of-plane gap closing converter at comparable pressures. At 0.01 atmospheres, simulations show less than 1 $\mu$W/cm$^3$ for the out-of-plane converter compared to 30 to 50 $\mu$W/cm$^3$ for in-plane gap closing converters. Finally, because limit stops can more easily be incorporated for in-plane operation, the in-plane gap closing converter does not suffer from the likely surface adhesion problems of the out-of-plane gap closing converter.

The model used in simulations for in-plane gap closing converters is developed in more detail in appendix B. An algorithm for the simulation incorporating the limit stops is also presented in appendix B.

## 5.    DESIGN OPTIMIZATION

Using the in-plane gap closing topology as the preferred concept, a more detailed design optimization can be done. The optimization was performed in Matlab with its built in functions that use a Sequential Quadratic Programming (SQP) method. The output of a dynamic simulation was used as the "objective function" for the optimization routine. The design variables over which the device can be optimized are shown in Table 6.1.

*Table –6.1*. Design variables for optimization.

| Variables | Description |
|-----------|-------------|
| $V_{in}$ | Input voltage |
| $l_T$ | Length of the shuttle (center) mass |
| $w_T$ | Width of the shuttle (center) mass |
| $t$ | Device layer thickness |
| $t_m$ | Proof mass thickness |
| $l_f$ | Length of the interdigitated fingers |
| $d$ | Nominal gap between fingers |

The width of the interdigitated fingers is determined by the combination of the device thickness ($t$) and the maximum aspect ratio constraint. Other parameters, such as proof mass and maximum capacitance are determined from these design parameters, design constraints, and a few assumptions.

The first constraint is that the total volume of the device must be less than 1cm³. As shown numerous times, the power conversion is linearly dependent on the proof mass. The mass is constrained by the total volume constraint and the material used. As silicon is a very lightweight material, it is not very desirable to use as a proof mass. Furthermore, the thickness of the device will not be more than 1mm, however, the total thickness of the device could be significantly greater given the volume constraint. The power output of the device can be greatly increased if an additional proof mass is attached to the silicon device. The resulting final device would look something like the model shown in Figure 6.11. As discussed in chapter 4, a tungsten alloy (90% tungsten, 6% nickel, 4% copper) makes a good proof mass because of its very high density (17 g/cm³). Therefore, the mass is constrained by the total volume constraint and the density of the tungsten alloy. A second constraint is the maximum aspect ratio achievable, which is about 50.

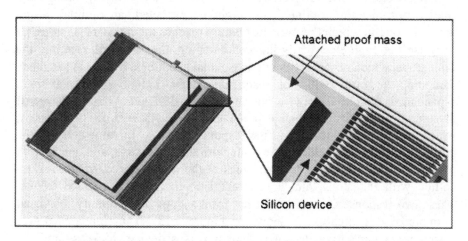

*Figure –6.11.* Model of an in-plane gap closing converter with proof mass.

Another assumption has been made which affects the design. Mechanical stops are designed to prevent the capacitive electrodes (the interdigitated fingers) from touching. The smaller the minimum dielectric gap (the closer the fingers are allowed to get) the higher the maximum capacitance. A realistic limit must be put on the minimum dielectric gap. Two optimizations were performed, one with a minimum dielectric gap of 0.25 μm and another with a minimum gap of 0.5 μm.

The optimization problem can then be formulated as shown in Figure 6.12. There are two nonlinear constraints. The first is the overall volume constraint, and the second arises from the aspect ratio constraint (50 in this case) imposed by the fabrication process. Remember that the other constraint of maximum aspect ratio also comes into play in the determination of the width of the fingers.

$$\text{Maximize:}\quad P = f\left(V_{in}, l_m, w_m, t, l_f, d\right)$$

$$\text{Subject to:}\quad \left(l_f + w_m\right)l_m t_m \leq 1cm^3$$

$$\frac{t}{d} \leq AspectRatio$$

$$V_{in}, l_m, w_m, t, l_f, d \geq 0$$

*Figure –6.12.* Formulation of optimization problem.

An interesting difficulty arises when running this problem through the Matlab optimization routine. As might be expected, an optimal design would be one in which the mass just barely reaches the mechanical stops. If there isn't enough electrically induced damping, the mass will ram into the limit stops, which obviously is not an optimal situation. If there is too much damping, the mass will not reach the stops, reducing the maximum capacitance, which is likewise not an optimal situation. The fundamental dynamics of the system change if the mass collides with the limit stops, which causes a discontinuity in the response surface. In some areas of the design space, the mass does not collide with the limit stops, and one smooth response surface results. In other areas of the design space, the mass does collide with the stops, and a different smooth response surface results. These two response surfaces meet, but form a slope discontinuity. A visual example of this situation is shown in Figure 6.13. The simulated power output for a capacitive design is shown versus device thickness and the nominal gap between fingers. The two different surfaces are clearly evident in the figure. The light portion of the response surface results when the shuttle mass does collides with the limit stops. The dark portion results when the shuttle mass does not reach the limit stops. Along the "ridge", the shuttle mass just barely reaches the limit stops. The optimization routine cannot calculate accurate first and second derivatives at the discontinuity. Since optimal designs will be very close the junction of these surfaces, if the optimization routine is run blindly, sub-optimal designs result. The optimization routine needs to be run in an iterative fashion with some engineering judgment. The optimization routine is run in limited design spaces where the discontinuity will not cause a problem. The output of these routines can be used with some intuition about the design space to select small areas that will be close to optimal, which would contain a discontinuity. With some judgment, the routine can be run on these smaller selected areas, and a close to optimal design can be chosen. It is very difficult to guarantee optimality with this approach, but practical designs that are nearly optimal can be generated.

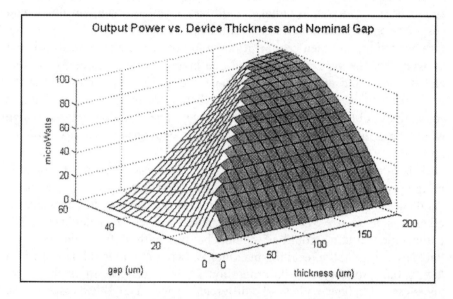

*Figure –6.13*. Simulated power output vs. device thickness and nominal gap between fingers. In the portion of the design space depicted by the light colored surface, the shuttle mass collides with the limit stops. In the dark portion, the shuttle mass does not reach the limit stops.

The resulting optimal design parameters, and the simulated power output, generated by the optimization routine are shown in Table 6.2. The optimal power output for a minimum dielectric gap of 0.25 µm is 116 µW and for a minimum gap of 0.5 µm is 101 µW. While these values are higher than those predicted by designs just based on engineering intuition by a factor of about 2 or 3 (see simulation results in previous section), they are still a factor of 2 or so lower than optimal designs for piezoelectric converters.

*Table –6.2*. Optimal design parameters and power output for an in-plane gap closing design.

| Variables | Description of Variable | 0.25 µm min gap | 0.5 µm min gap |
|---|---|---|---|
| $V_{in}$ | Input voltage | 10 V | 10 V |
| $l_T$ | Length of shuttle mass | 9 mm | 8 mm |
| $w_T$ | Width of shuttle mass | 10 mm | 10 mm |
| t | Device layer thickness | 200 µm | 200 µm |

| Variables | Description of Variable | 0.25 µm min gap | 0.5 µm min gap |
|---|---|---|---|
| $t_m$ | Proof mass thickness | 5 mm | 5 mm |
| $l_f$ | Length of fingers | 50 µm | 50 µm |
| d | Nominal dielectric gap | 530 µm | 1.2 mm |
| Pout | Output power | 116 µW | 101 µW |

An input voltage of 10 volts may be quite high for wireless sensor applications. The output voltage could be converted down, but this would cost extra power. However, it may still be preferable to use a higher voltage for conversion, and then convert it down to power the electronics. It is not unlikely that the system would need one higher voltage source for sensors and actuators and another lower voltage source for electronics. With the load electronics and power circuitry better defined, a more limiting constraint could be put on the input voltage. The resulting optimal design would be a little lower in terms of power conversion, but may still be preferable.

It is often the case that the device layer thickness will not be a flexible design parameter. For example, if a standard foundry process is to be used, the device thickness will be set by the process. It also may be necessary for other reasons to use SOI wafers with another device layer thickness. The optimization routine could of course be run with a fixed device layer thickness. Again, the resulting maximum power output would be somewhat lower, but probably not in proportion to the decrease in device layer thickness. Designs with vastly different device layer thicknesses could, therefore, still be attractive.

## 6. FLEXURE DESIGN

The design of the flexures is left out of the optimization routine. The flexures must satisfy four demands:

1. The natural frequency of the device should match that of the input vibrations. Because the system mass is determined by the already obtained design parameters, the flexures must have a predetermined stiffness.
2. Given the designed range of motion along the axis of the driving vibrations, the fracture strain of the springs should not be exceeded (some factor of safety should be designed in), and the springs should ideally remain in the linear region.
3. The springs should be stiff enough that the static deflection out of the plane of the wafer should be minimal, significantly less than the space

between the device and the substrate. Also, the strength of the springs should be strong enough in the out-of-plane direction that the force of gravity on the large proof mass will not cause them to fracture.
4. The flexures should be significantly less stiff along the desired axis of motion compared with the out-of-axis directions.

The layout of a device is shown in Figure 6.14. Fairly standard folded flexures are used at each of the corners as shown in the figure. The overall lateral (in-plane) stiffness of the flexures is given by equation 6.19 and the vertical (out-of-plane) stiffness is given by equation 6.20.

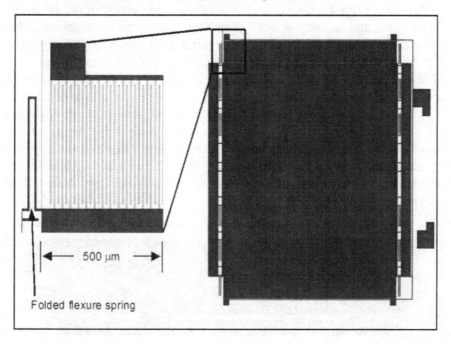

*Figure –6.14.* Layout of an in-plane gap closing converter with a close-up of the folded flexure spring at the corner.

$$k_l = \frac{NEtw_{sp}^3}{2l_{sp}^3} \tag{6.19}$$

$$k_v = \frac{NEw_{sp}t^3}{2l_{sp}^3} \tag{6.20}$$

where $N$ is the number of folded flexures (springs), $E$ is the elastic (Young's) modulus of the material (silicon in this case), $t$ is the device thickness, $w_{sp}$ is the width of the flexure beams, and $l_{sp}$ is the length of the flexure beams.

The device thickness will generally already be determined by the optimization routine, and the elastic modulus is a fixed material property. Therefore, only the spring length, width, and number of springs can be altered to achieve the desired stiffness. Incidentally, more than four folded flexures could be used. Actually, a total of twelve flexures are used on the device shown in Figure 6.14. A closer image of one side of this device, with six flexures, is shown in Figure 6.15. Note that the device has been rotated 90 degrees in this figure. It is also possible to include multiple folds in a flexure structure to make it more compliant. However, in practice this has not been necessary because of the large proof mass attached. It is important that the vertical stiffness be about 10 times the lateral stiffness to reduce out-of-axis motion. The static vertical deflection under the weight of the proof mass is also important. The static deflection is simply the gravitational force over the vertical stiffness ($mg/k_v$). The length and number of springs affect the vertical and lateral stiffness in exactly the same way. However, the width of the springs is linearly related to the vertical stiffness and but related to lateral stiffness by the third power. So, one can more or less set the desired vertical to lateral stiffness ratio be correctly selecting the width of the springs.

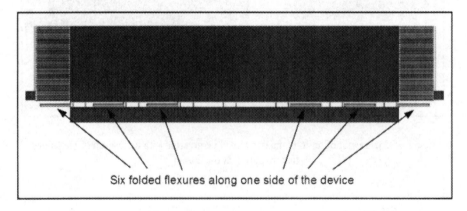

Six folded flexures along one side of the device

*Figure –6.15.* Layout of one side of a capacitive converter device showing the six flexures in parallel to increase the stiffness.

The maximum lateral deflection is determined by the nominal dielectric gap minus the minimum dielectric gap. For the design shown in Table 6.2 the maximum lateral deflection is 49.75 µm. The maximum stress due to the lateral deflection is given by equation 6.21. The fracture stress for single crystal silicon is about 70 GPa. The stress value calculated by equation 6.21 does not take into account stress concentrations. A suitable factor of safety needs to be chosen, and then the maximum stress must then remain below the yield stress divided by the factor of safety.

$$\sigma_l = \frac{3Ew_{sp}x_{max}}{2l_{sp}^2} \tag{6.21}$$

where $x_{max}$ is the maximum lateral deflection.

It is also desirable, although not absolutely necessary, that the flexures remain in the linear region under the maximum deflection. For the given type of flexures (2 link beams), a rough rule of thumb for linearity is that the maximum deflection divided by the length of the flexure (that is one link of the flexure) be 0.5 or less. Linearity depends on the assumption that for the maximum slope or angle ($\theta$) in the beam, tan($\theta$) is approximately equal to $\theta$. For the rule of thumb given above, tan($\theta$) equals 1.013 times $\theta$, or a 1.3% error. For the design shown in Table 6.2, the flexure of the beams needs to be less than 100 μm to remain in the linear region.

The maximum stress due to the static vertical loading of the proof mass is given by equation 6.22. As with the lateral strain, the flexures must be designed such that this stress is below the yield stress divided by a suitable factor of safety. It is likely that there will be some dynamic loading in the out-of-plane direction. Vertical limit stops of some sort would need to be incorporated in the package limiting the maximum vertical displacement to a specified value ($y_{max}$). The maximum stress in the flexures as a function of the maximum allowed vertical displacement is given in equation 6.23. Again, the system should be designed such that this vertical displacement will not result in a fractured flexure. If the designer needs to improve the factor of safety and/or the linearity of the flexures, the length and the number of the flexures can both be increased such that the lateral and vertical stiffness will not change, but the maximum developed stress will decrease, and the ratio of lateral deflection to spring length will decrease, thus improving the linearity.

$$\sigma_v = \frac{3mgl_{sp}}{Nw_{sp}t^2} \tag{6.22}$$

where $m$ is the proof mass, and $g$ is the gravitational constant.

$$\sigma_v = \frac{3Ety_{max}}{2l_{sp}^2} \tag{6.23}$$

Given the relationships presented here, the flexures should be designed to meet all four of the criteria stated above. A short example of the design of the flexures for the design shown in Table 6.2 follows. The specified device thickness is 200 μm. The desired natural frequency is 120 Hz. The proof mass is calculated at 7.4 grams. The necessary lateral stiffness then becomes 4.2 kN/m. There are three parameters to specify to get the desired stiffness: beam width, beam length, and number of springs. The three must be selected in somewhat of an iterative process to make sure that acceptable factors of safety are met, and acceptable linearity is maintained. After iterating a few times and making calculations (an algorithm for this purpose is easily developed in a package like Matlab), twelve was selected as a suitable number of springs. The line shown in Figure 6.16 shows acceptable values for the width and length of the beams making up the 12 springs.

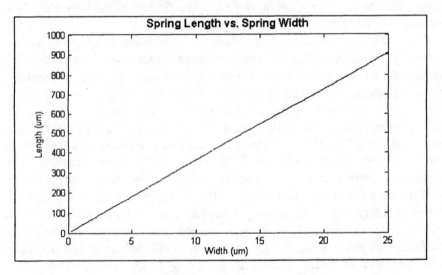

*Figure –6.16.* Acceptable spring flexure lengths and widths with 12 springs in parallel in order to generate a natural frequency of 120 Hz.

A width of 20 μm and the corresponding length of 727 μm were selected as an acceptable design. Again these values were selected based on calculations of factors of safety, linearity, and static vertical deflection. Given a maximum lateral deflection of 49.75 μm, the factor of safety along the axis of motion is 14.7. Stress concentrations have not been considered which is why such a high factor of safety is desired. If it was necessary to make the beams shorter and thinner, stress concentrations could be taken into account when calculating the maximum stress, and then a much smaller factor of safety could safely be used. The ratio of maximum lateral displacement to beam length is 0.07, which is very safely in the linear region. The vertical to lateral stiffness ratio is a very comfortable 100. The static deflection in the vertical direction is 0.2 µm, and the factor of safety based on this static deflection is a huge 423. If a reasonable minimum factor of safety of 5 in the vertical direction were desired, then the resulting maximum allowable vertical deflection would be 15 µm. Fabricating limit stops to limit the vertical deflection to 15 µm could be a significant challenge. These flexures are somewhat over-designed, however given the huge size of the overall device (by MEMS standards), there is plenty of physical space for them.

# 7.     DISCUSSION OF DESIGN AND CONCLUSIONS

Perhaps the most important point to be made about this design is that the power density of an optimized design is at least a factor of 2 lower than the optimal power from a piezoelectric design. As discussed previously in chapter 3, the energy than can be converted by electrostatic transducers is inherently lower than piezoelectric converters. While there is no claim that the design topology and method used cannot be improved, it is very unlikely that any electrostatic design can be generated that can match a good piezoelectric design at the meso-scale.

The power converted is unavoidably linked to the mass of the system. Given this fact, along with an overall volume constraint of 1cm$^3$, the resulting design is extremely large by MEMS standards. Also, unfortunately, in order for the electrostatic converter to generate a significant amount of power, an additional proof mass needs to be attached to the device. The shear size of the device negates one of the potential advantages of MEMS technology, which is low cost due highly parallel manufacturing methods. Additionally, given the large size, integration with microelectronics is less useful. Furthermore, the adding of such a large mass to a device with micron sized features results in a very delicate device, which would likely not be very robust.

The real advantage of a MEMS based electrostatic converter is only realized if the device is considerably scaled down in size. The dynamic models, design procedures and principles presented in this chapter will, of course, still apply to vastly smaller designs. However, the potential power conversion from the smaller designs is also much smaller because of the lower system mass. The scaling of power conversion goes down linearly with the system mass (and therefore the system volume). A converter of size 1 mm$^3$ would then have a potential power density of about 100 nW based on the input vibrations currently under consideration. It is possible that applications with vastly higher energy vibrations, and therefore larger power output per unit volume, could make much smaller designs attractive. However, whether designs on the order of 1mm$^3$ will ever be very useful is an open question.

Chapter 7

# FABRICATION OF ELECTROSTATIC CONVERTERS

The design of electrostatic converters has been discussed in the previous chapter. Perhaps more than with other manufacturing technologies, the design of MEMS devices is closely tied to the target processing technology. Therefore, a very short discussion of the manufacturing process to be used has been given in chapter 6 in order to effectively produce a design. The processing that has been used will be covered in more depth in this chapter. A number of variants on the basic SOI process have been used to fabricate a string of prototypes. Each of these variants and the purpose for which the variant was used will be discussed.

## 1. CHOICE OF PROCESS AND WAFER TECHNOLOGY

As explained in chapter 6, an SOI wafer and processing technology have been selected due to the large device thickness, and therefore large capacitances that can be generated. Additionally, the Deep Reactive Ion Etching (DRIE) etching process used to create devices in the SOI wafer is capable of very high aspect ratios (up to about 50). This also improves the potential maximum capacitance. A further benefit of the very thick device layer and high aspect ratio is that the resulting devices have a very high out-of-plane stiffness compared to the in-plane stiffness. This is important in order to allow the addition of a significant amount of mass to the system after the processing is done. One potential drawback of the SOI technology is that it only provides one structural layer with which to design devices. However, only one layer is really needed for the current application.

Therefore, SOI MEMS technology seems best suited for the design of electrostatic vibration-to-electricity converters.

Over the past few years, it has become increasingly common to etch MEMS structures into the device layer of SOI wafers. However, although the technology exists, at the outset of this project there was no standardized process available to the public. Recently Cronos (Cronos 2003) has added an SOI process to their standard three-layer polysilicon process (MUMPS) and made it available to the public.

There are a couple of issues relating to the processing that will increase the complexity somewhat. First, it is essential to minimize parasitic capacitance. In typical MEMS devices, all of the electronics are on a separate die from the MEMS die. Electrical contact is usually made with wire bonds. In this case, at least two electronic devices need to be more closely integrated, the two switches that control the flow of charge into and out of the variable capacitor (see Figure 6.1). A second contributor to parasitic capacitance is the substrate beneath the device. It is therefore beneficial to etch away the substrate directly under the device. This backside etching is commonly done, but does increase the processing complexity. Finally, it is important that the single crystal silicon be highly conductive to reduce resistive losses as much as possible.

## 2.    BASIC PROCESS FLOW

Several slightly different processes have been used to create a sequence of prototype devices. These processes are all based on the same SOI MEMS technology, but differ in certain respects. This section presents a basic process flow that all of the processes more or less follow. The following section will present how each specific process used differs from the basic process flow, and why that process was used. It is assumed that the reader is familiar with micromachining processes and terminology, and so only high level explanations of the processes will be given here. The reader is referred to Jaeger (Jaeger 1993) and Madou (Madou 1997) for a detailed discussion of microfabrication.

Figure 7.1 shows the basic process flow as schematics of the cross section of the wafer at sequential stages of the process. The process begins with an SOI wafer, which is a sandwich of single crystal silicon, silicon dioxide, and single crystal silicon. The cross section is not to scale. The wafers used for this project had top silicon layers (the device layer) ranging from 7 μm to 50 μm thick, oxide layers of 2 μm thick, and bottom silicon layers (the substrate) of about 400 μm thick.

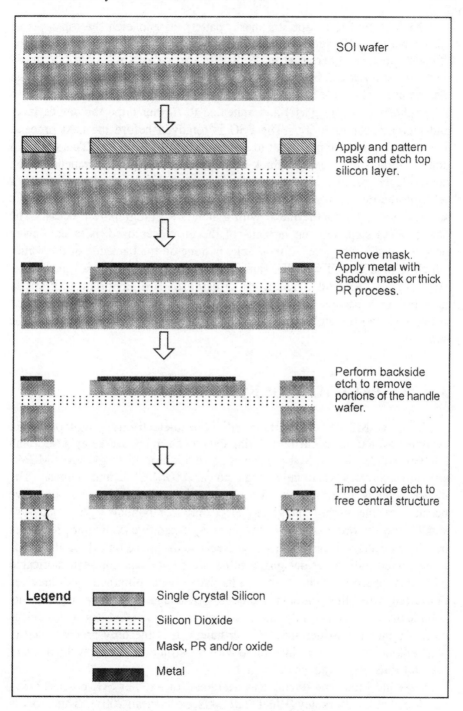

SOI wafer

Apply and pattern mask and etch top silicon layer.

Remove mask. Apply metal with shadow mask or thick PR process.

Perform backside etch to remove portions of the handle wafer.

Timed oxide etch to free central structure

**Legend**

Single Crystal Silicon

Silicon Dioxide

Mask, PR and/or oxide

Metal

*Figure –7.1.* Basic process sequence used to fabricate capacitive converters.

The first step is to apply a mask, pattern it, and etch the top layer of silicon with a DRIE process, most likely what is generally referred to as the "Bosch" process (Laerme *et al* 1999). In many cases a simple UV baked photo resist mask can be used. If necessary, a hard mask made of oxide can also be used. In some of the processes used, the wafer is then covered with PSG (phosphosilicate glass) and annealed to further dope the device layer and increase conductivity. The PSG is removed before the next process. The next step is to apply metal to the top layer of the wafer. This has been done in three ways: using a thick resist process to fill in the trenches in the device layer, using a shadow mask (Cronos 2003) to apply the metal, or refilling the device layer trenches with oxide to more or less planarize the surface and applying the metal with a standard lithographic process. The next process step, left out in some of the processes used, is to etch away portions of the substrate. Lithography is done on the backside of the wafer, which is then etched with the same DRIE process. Finally, a timed oxide etch removes the oxide, freeing the structures. It is important to properly time this etch because unlike most surface micromachined devices, oxide remaining after the etch forms the anchors between the device layer and the substrate.

## 3.    SPECIFIC PROCESSES USED

The first MEMS electrostatic vibration-to-electricity prototypes were designed for a process currently being developed in UC Berkeley's Microlab (Bellew 2002). It is a SOI process that, in addition to single layer MEMS structures, creates solar cells, high power MOSFETS, and diodes. The device layer is 15 μm thick. The process is very complex, but attractive because of the highly integrated electronic components. A number of smaller (on the order of $1 - 4$ mm$^2$ in area) capacitive converter prototypes have been designed for this process. Because the space on a 1cm$^2$ die needs to be shared with other designs, a full sized device has not been fabricated with this process. The best results have been obtained with devices fabricated with this process. The disadvantage of this process is its complexity. It is actually more complex than needed for the devices currently under consideration. Nevertheless, it is the only process used in which diodes can be fully integrated with the structure, which is the primary reason for its greater success.

A second prototype device was designed for another SOI process being developed at UC Berkeley (Rhiel *et al* 2002, Srinavasan 2001). This process is much like the basic one shown in the previous section except that the backside etch is not performed. After the processing sequence is shown, a fluidic self-assembly process is used to assembly bare die FETS or diodes directly to the silicon die. The electronic components are not as highly integrated as in the process just described, however, this is a simpler process. Additionally, the bare die FETS assembled are of better quality than those fabricated in Bellew's process. Although the method of attachment for bare die diodes was verified, the mechanical structures fabricated were not functional. The structural layer was only 7 μm thick and so the capacitances generated by the variable capacitance structure were not very high. It seems that the parasitic capacitances dominated the system, and so the system did not function well. Additionally, no method of manually actuating the structures was included in the design. So when the structures did not work under vibration, it was very difficult to troubleshoot the problem because the structure could not be actuated any other way.

A third set of prototypes was fabricated by the authors with a simplified process identical to that shown in Figure 7.1. The device layer of the wafers used with this process was 50 μm thick. Two methods of electrically connecting to the diodes were used. First, small packaged surface mount diodes were used and wire bonds provided the connection between the MEMS die and the diodes. Second, bare die diodes were attached to the MEMS die manually with conductive epoxy. The purpose of running this process was first to build devices with a thicker structural layer thus increasing the maximum capacitances, and to have the capability of building larger devices to which a tungsten proof mass could be attached. Tungsten proof masses were attached to the top surface of the devices manually with epoxy.

While the fabrication of the third set of prototypes was being performed, Cronos announced a new SOI MEMS process being made available to the public (Cronos 2003). The authors had the good fortune of being able to submit a design for a trial run on this process. This new process is a three mask process and is performed almost exactly as is shown in Figure 7.1. The metal is applied with a shadow mask, and the device layer for the trial run was 25 μm. The intent was to manually assemble bare die diodes with conductive epoxy on the die. The prototype device submitted to this process was non-functional when it arrived. The large center plate or shuttle mass was completely broken off during processing. After the trial run, Cronos changed the thickness of the device layer to 10 μm, which limits the usefulness of this process for electrostatic vibration-to-electricity converters because of the lower capacitances achieved with the thinner device layer.

While each of these processes is a little different, they are all alike in one key respect; *they all create a micromechanical device from a relatively thick device layer on an SOI wafer.* Results from prototypes fabricated with each of these processes are presented in the following chapter.

## 4.    CONCLUSIONS

An attempt was made, initially, to either use standard publicly available processes or "piggy-back" on other processes currently being run in UC Berkeley's microlab for a couple of reasons. The first, and most important, is that the innovative and important contribution of this project is in the design and modeling, not in the processing, of the device. Therefore, if a suitable, available process could be found, time and money could be saved, and the likelihood of success would also be increased. A second reason for trying to find a standard process to use is that the scope of this thesis is quite broad, including the detailed modeling, design, and construction of both piezoelectric and electrostatic generators. It was felt that if a too many resources were committed to developing a process, the other portions of the project would suffer. However, suitable commercial processes could not be found. Therefore, an attempt was made to use processes currently under development in the UC Berkeley Microlab. This however, was less than completely successful. Finally, it was decided to run a custom, but simplified, process in order to be able to fabricate devices to demonstrate the electrostatic vibration-to-electricity conversion concept. More development work needs to be done on this custom process in order for it to be a viable and reliable process for developing electrostatic vibration-to-electricity converters.

# Chapter 8

# ELECTROSTATIC CONVERTER TEST RESULTS

A sequence of prototype devices has been designed and fabricated. The previous chapter explained the essential elements of the fabrication processes used. This chapter is dedicated to reporting fabrication and testing results.

## 1.    MACRO-SCALE PROTOTYPE AND RESULTS

A macro-scale prototype was first built using more conventional machining processes in order to verify the basic concept of operation and the test circuit before investing resources into the fabrication of MEMS devices. This device is shown in Figure 8.1. The basic device is comprised of three parts. An aluminum piece, 500 μm thick, was milled on a high spindle speed three axis machine using a milling cutter 229 μm in diameter. This piece, shown alone in Figure 8.2, serves as a flexure and an electrode. (One of the flexures is broken on the part shown in Figure 8.2.) The slots cut through the piece, which form the flexural structures, are 229 μm in width. A pocket is cut out from the backside so that there will be a dielectric gap of 250 μm in the nominal position. A printed circuit board was etched to form the other electrode of the variable capacitor and the electrical connection lines for the circuit element. This forms the bottom layer of the structure and is clearly visible in Figure 8.1. Finally, a steel mass is attached to the top of the aluminum piece. The device is an out-of-plane gap closing converter. Although this is not optimal for MEMS design, it was the most practical configuration for a macro scale test prototype. The copper electrode of on the circuit board under the aluminum piece was covered with a thin dielectric (the ink from a Sharpie felt tip pen) in order to ensure that the two electrodes from the variable capacitor did not make electrical contact.

*Figure –8.1.* Photograph of the first macro-scale prototype built.

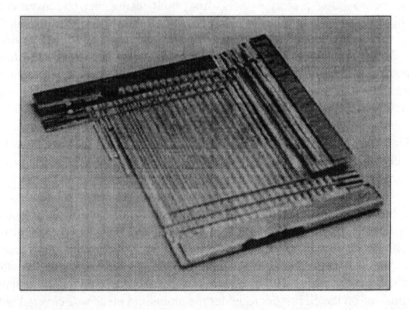

*Figure –8.2.* Photo of the backside of the aluminum piece used in the macro-scale prototype.

The circuit shown in Figure 8.3 was used to measure the output of the macro-scale prototype. The op-amp in a unity gain buffer configuration was included in the circuit in order to decouple the capacitance of the storage capacitor from that of the measurement system. The output of this circuit was then measured on a standard oscilloscope.

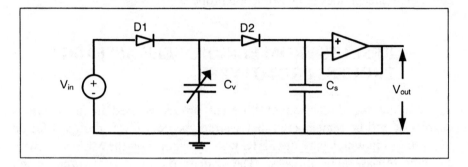

*Figure –8.3.* Measurement circuit for the capacitive converter prototypes.

Two surface mount diodes and a very small ceramic capacitor are visible in Figure 8.1. The choice of diodes is very important. A few iterations with different diodes were performed before the system worked properly. First, the capacitance of the diodes will add to the parasitic capacitance that the variable capacitor sees. So, the capacitance of the diodes should be as small as possible. Secondly, the reverse leakage current varies considerably for different types of diodes. Reverse leakage currents generally range from about 0.1 nA to about 100 μA. An intuitive explanation of the problem is as follows. On a capacitor of 100 pF (about the size of interest in this case), a 0.1 nA leakage will result in a voltage drop of about 1 V/s, which is OK given that the circuit is operating at about 100 Hz. However, at 1 μA leakage, the voltage would drop at 1000 V/s, which is clearly too fast. The problem is that as the capacitance of $C_v$ decreases (and its voltage should increase), current will be flowing back through the input diode, *D1*, so fast that the voltage will never increase at the output. Furthermore, if the output voltage across $C_s$ were to increase, the current would flow back onto $C_v$ too fast to detect a change. Initially diodes with a low forward drop were chosen, however, these did not work well because their reverse leakage was too high. In general, diodes with a lower forward drop also have a higher reverse leakage current. It is therefore preferable to choose a diode with low leakage current and high forward drop. The diodes chosen have a leakage current of 10 nA as stated by the manufacturer.

The converter was tested by manually pushing the proof mass up and down. With a storage capacitance of 100 pF and a source voltage of 3 volts, the output voltage increased 0.25 volts per cycle. With a source voltage of 9 volts, the output voltage increased 1 volt per cycle. At an operation frequency of 100 Hz, the output power would then be less than 0.1 nW for a 3 volt source and about 1 nW for a 9 volt source. This is not very good power production, but at least the device demonstrates that the capacitive converter concept does in fact function properly.

## 2. RESULTS FROM FLUIDIC SELF-ASSEMBLY PROCESS PROTOTYPES

Recalling the discussion of different processes used in the previous chapter, it will be remembered that a prototype device was designed for an SOI process in which bare die FETS were assembled on the MEMS die with a fluidic self-assembly process. The primary purpose of this process and these prototypes was to verify that either bare die FETS or diodes could be assembled onto the MEMS die and function effectively as switches (or diodes). A scanning electron micrograph (SEM) of the device fabricated is shown in Figure 8.4.

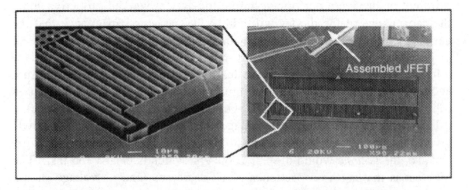

*Figure –8.4.* SEM of prototype device fabricated with fluidic self-assembly process.

The bare die JFETS that were assembled are clearly visible in the lower magnification picture on the right. The JFETS were wired to act as diodes as shown in Figure 8.5. Bare die JFETS were used in place of diodes primarily because of availability, but also because they tend to have low reverse leakage when wired as diodes. The JFETS were tested after assembly. One of the I-V curves from the tests is shown in Figure 8.6. Figure 8.6 shows the same data plotted on two different scales. The left graph is a linear scale. The right graph plots the absolute value of the data on a log scale. A log scale is used so that the reverse leakage current can be seen more precisely. The reverse leakage current is about 0.2 nA, which is very good compared to commercially available diodes. The forward voltage drop is about 0.75 volts, which is quite high. However, as mentioned before, there is a tradeoff between leakage current and forward voltage drop, and for this application is better to have a low leakage and a high voltage drop. Therefore, the JFETS wired as diodes seem to be a good choice for this process.

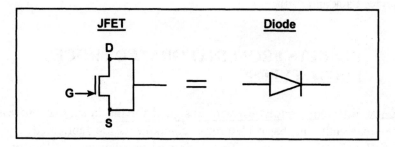

*Figure –8.5.* Illustration of a JFET wired to operate as a diode.

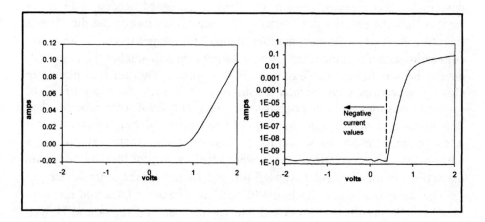

*Figure –8.6.* Current vs. voltage curve measured from a fluidically self assembled JFET wired as a diode. The left graph shows data on a linear scale, the right graph shows the absolute value of the data on a log scale.

The size of the variable capacitance device was only about 300 μm X 1mm due to space constraints on the die. Furthermore, the device thickness was only 7μm. It is doubtful whether a device of this size and thickness would be able to generate a large enough capacitance to overcome the parasitics. In tests, the structure failed to increase the voltage on the output capacitor when driven by vibrations. As explained in chapter 7, it is believed that this is because the small capacitance of the variable capacitance structure as compared to the parasitic capacitances. There was one more problem noted with the design of this device. It is useful to be able to manually move the variable capacitor back with a probe tip or by some other method in order to verify the correct operation of the system before putting it on the vibrometer. However, no probe access points were designed in this particular device, so it was not possible to manually move the device back and forth.

## 3.    RESULTS FROM INTEGRATED PROCESS PROTOTYPES

A few different prototypes were designed for the integrated process in which solar cells and basic electronic components are fabricated together with SOI MEMS structures. The design of the devices for this process happened after the prototypes from the fluidic self-assembly process, so lessons learned from that prototype were incorporated in the design of devices for this process. Again, the devices designed needed to be much smaller than the optimal size because of space constraints on the die. It was therefore decided to design a number of small prototypes to verify the power conversion concept rather than a single larger (but still smaller than optimal) device that would be capable of more power output. Two small in-plane gap closing converters with thermal actuators next to them for the purpose of manually pushing the converter devices back and forth were fabricated in this process. It was thought that it would be easier to debug the converter if it could be actuated in a controlled fashion rather than just put on a vibrometer. The converter devices were identical except that the minimum dielectric gap for one device was 0.5 µm and 1.0 µm for the other device.

The thermal actuators designed to push the converter back and forth are based on the design first proposed and carried out by Cragun and Howell (Cragun and Howell 1999). This actuator is referred to as a Thermal In-plane Microactuator or TIM. Figure 8.7 shows a schematic of a TIM and an associated converter. It should noted that the converters in this case are fabricated such that the movable fingers are offset in the nominal position so that as voltage is put across the variable capacitor, the electrostatic forces will cause it to naturally close, and then the TIM actuator will modulate the dielectric gap. Current is passed through the many arms, which heat up as a result of the power dissipated. As the arms heat up, they expand pushing the center yoke forward. Thermal actuators generally give a high force, low displacement output compared to electrostatic actuators. The TIMs are attractive in that they generate high force at zero displacement, which is necessary to overcome the electrostatic force pulling the converter combs together, but also can produce displacements on the order of 10 μm. Models and results have been presented for TIMs surface micromachined from polysilicon, but the author is not aware of any TIMs being implemented in an SOI process. The TIMs were carefully modeled and designed to provide both adequate force and displacement output for the associated converter. However, detailed discussion of these models and designs is beyond the scope of this book. The reader is referred to Lott *et al* for detailed discussion of the modeling of polysilicon TIMs (Lott 2001, Lott *et al* 2002).

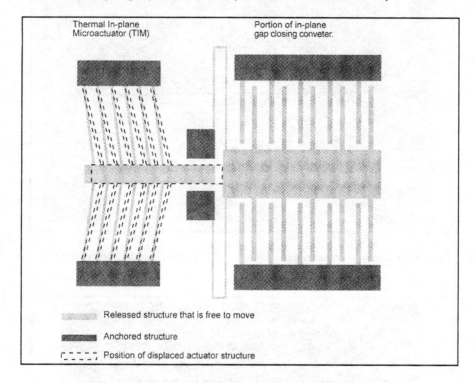

*Figure –8.7.* TIM actuator designed to push converter structure back and forth.

Figure 8.8 shows a Scanning Electron Migrogaph (SEM) of a TIM and converter structure fabricated with this process. Figure 8.9 shows a close-up view of the interdigitated comb fingers and spring flexure. Figure 8.10 shows a sequence of images from a normal optical microscope. The sequence, going from left to right, shows the converter and TIM in their nominal positions, partly actuated or extended, and fully actuated. As the converter is pushed from its maximum capacitance position to its minimum capacitance position the voltage across the output capacitor should increase, thus verifying the correct operation of the system. Note that Figure 8.10

actually shows the converter being pushed past its minimum capacitance position. The middle image shows the minimum capacitance position while the images on the right and left show higher capacitance positions. The image sequence shows the converter being pushed past the minimum capacitance position so that the motion can more easily be seen.

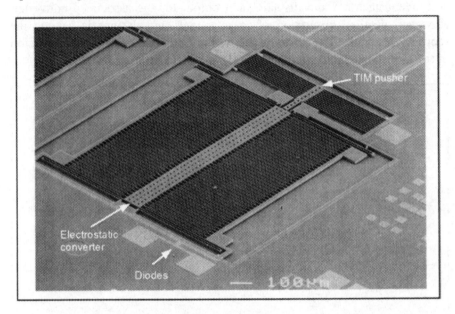

*Figure –8.8.* SEM of electrostatic converter with TIM pusher from the integrated process.

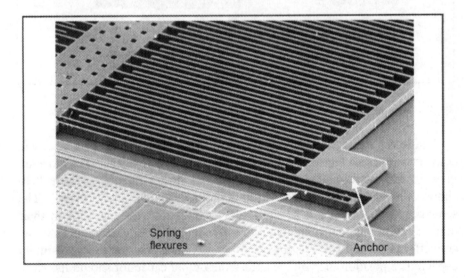

*Figure –8.9.* Close-up of interdigitated fingers and spring flexures.

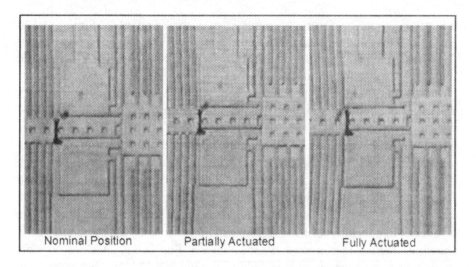

*Figure –8.10.* Image sequence showing the TIM actuator pushing the converter from its nominal position to a fully extended position.

Two devices were successfully tested with an input voltage of 5 volts and an output capacitance of 100 pF. As the TIM actuator pushed the converter from a high capacitance position to a low capacitance position, the output voltage across the 100 pF capacitor increased by 0.3 volts. Based on geometry, the parasitic capacitance should be about 4.2 pF. It is, however, likely that the real parasitic capacitance is somewhat more than that. The maximum and minimum capacitances of the converter structure were calculated as 9.6 pF and 1.2 pF respectively. Using these parameters, the calculated voltage gain at the output per cycle should be 0.4 volts. The voltage across the output capacitor fed the input to an operational amplifier used in a unity gain configuration as shown above in Figure 8.3 in order to decouple that parasitic capacitance of the measurement equipment from the output capacitor. However, some parasitic capacitance at the output still exists which may have a significant effect on the voltage gain per cycle. For example, if the output capacitance were really 125 pF, the calculated voltage gain would only be 0.32 volts. The energy gain associated with the 0.3 volt increase is 1.4 nJ. Assuming a driving frequency of 120 Hz, and remembering that this device undergoes two electrical cycles for each mechanical cycle, the power output would be 337 nW. The size of the entire converter device is about 1.2 mm X 0.9 mm X 0.5 mm (including the thickness of the substrate). Based on this volume, the power density would then be 624 $\mu$W/cm$^3$. However, this assumes that the input vibrations would be able to mechanically drive the structure hard enough to overcome the electrostatic forces. Given the very small mass of this system, such vibrations would be far more energetic than the standard 2.25 m/s$^2$ at 120 Hz. Nevertheless, the structures fabricated with this integrated process did demonstrate the basic functionality of a micromachined electrostatic vibration-to-electricity converter.

Although the design and modeling of the TIM structures has not been discussed, a few results may be of interest for future researchers. Two TIM pushers were designed. The first was capable of generating higher forces with lower displacements. This structure consisted of 32 beams (16 on each side of the yoke), which were 7.5 μm wide, 15 μm thick, and 400 μm long. The resistance of the structure was 1 kΩ. The second pusher was designed for smaller forces and higher displacements. It consisted of 24 beams (12 on each side of the yoke), which were 5 μm wide and 400 μm long. The resistance of this structure was 2 kΩ. It is this second TIM structure that is shown on the entire converter device above in Figure 8.8. A close-up view of this TIM structure is shown below in Figure 8.11. In both cases, the beams were angled from the horizontal by 3 μm. So the angle of inclination was sin$^{-1}$(3 / 400) = 0.43 degrees. The resistivity of the silicon material can be back calculated from the resistances of the TIMS as 0.23 Ω-cm.

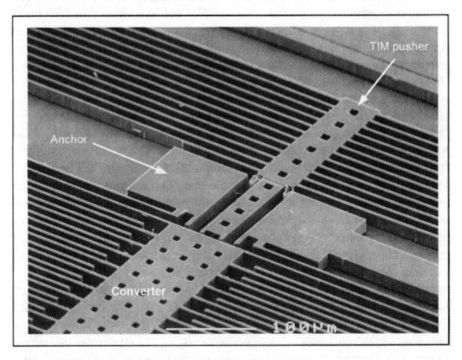

*Figure –8.11.* Close-up of TIM and yoke, which pushes the electrostatic converter.

Both structures were designed to reach a maximum displacement with an actuating voltage of 30 volts. The yoke began to move when the voltage across the TIM was about 7 volts, and the structure reached its maximum displacement at a voltage of about 28 volts for both structures. Additionally, in both cases, the structure broke at a voltage of about 32 volts. The second TIM, designed for less force and higher displacement, was not able to produce enough force to overcome the electrostatic attraction of the comb fingers and move the converter structure with an input of 10 volts to the electrostatic converter. Figure 8.12 shows an optical microscope image of the TIM structure pushing on the electrostatic converter with an input voltage of 10 volts. Notice that the beams on the TIM structure are starting to buckle under the force and are still not able to move the converter structure. However, with an input of 5 volts, the TIM could easily push the converter structure. In fact, the higher displacement TIM actually pushed the converter structure well past its minimum capacitance position as shown in Figure 8.10. The higher force, lower displacement, TIM could move the converter structure with an input of 10 volts, but could not push the converter past its minimum capacitance position.

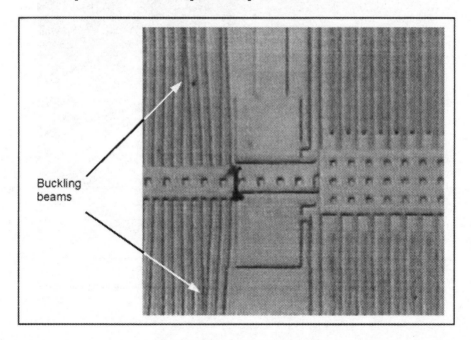

Buckling beams

*Figure –8.12.* TIM structure trying to push electrostatic converter with an input of 10 volts, causing the beams on the TIM to buckle.

Another very important, perhaps even the most important, element of the integrated process designs is that the diodes acting as the input and output switches are fabricated into the wafer right next to the electrostatic converter structure. Thus, the parasitic capacitances are minimal and the characteristics of the diode can be controlled to a certain extent. Figure 8.13 shows a close-up image of the diode structures next to the converter structure. Figure 8.14 shows an I-V curve resulting from testing on one of these diodes. As above in Figure 8.6, the values for the current are shown both on a linear scale and on a log scale (absolute values shown). As mentioned previously, the critical parameter in this case is the reverse leakage current. The average reverse leakage current as shown in Figure 8.14 for the integrated diodes is about 50 nA, which is not nearly as good as the 0.2 nA from the assembled bare die JFETS, but still adequate.

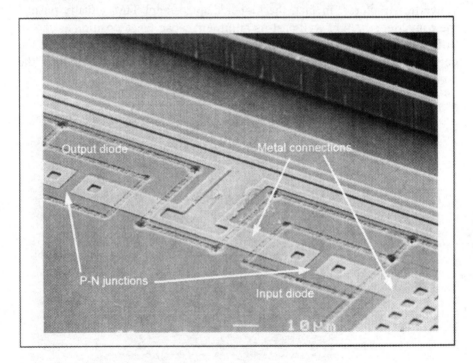

*Figure –8.13.* Close-up view of input and output diodes fabricated in the integrated process.

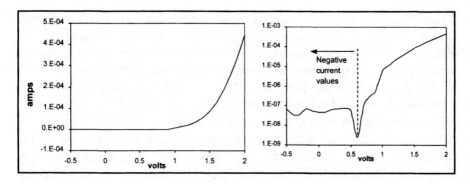

*Figure –8.14.* I-V curve generated from one of the diodes shown in Figure 8.13. Data are shown on a linear scale on the left graph, and the absolute value of the data is shown on a log scale on the right graph.

## 4. RESULTS FROM SIMPLIFIED CUSTOM PROCESS PROTOTYPES

As discussed in chapter 7, the authors also developed and ran a simplified custom process for basically two reasons: first, in order to have enough space to design and fabricate full sized devices, and second, to fabricate devices of greater thickness (50 μm) increasing the maximum capacitance of the devices. The masks for this run covered the entire wafer rather than a 1cm X 1cm die. (i.e. A stepper was not used for lithography, rather a contact lithography method was used.). Additionally, wafer space was not shared with other users. The result was that many larger designs could be accommodated. Optimal designs with a total area constraint of 1 cm² and 0.25 cm², a thickness constraint of 50 μm, and minimum dielectric gaps of both 0.25 μm and 0.5 μm were fabricated. Additionally, test devices incorporating the TIM "pushers" were designed and fabricated.

One of the larger converters fabricated by this process is shown in Figure 8.15. The device shown is a 0.25 cm² device. A close-up view of the interdigitated comb fingers and a spring flexure is shown in Figure 8.16. The very large center plate is meant to accommodate additional mass to be attached after the process. More than four spring flexures (one at each corner) need to be included on this device in order to achieve the correct stiffness to produce a resonant frequency of 120 Hz with the additional mass. These extra spring flexures are barely visible, but are pointed out, in Figure 8.15. Because the oxide below the structural layer is used to anchor the structure to the substrate beneath, only about 10 μm of oxide are etched away under the structure. The anchors then must be significantly larger than

20 μm X 20 μm so that enough oxide remains under them to act as a good anchor. However, the 10 μm oxide etch must release the enormous center plate, therefore a huge array of etch holes meant to allow the hydrofluoric acid to go beneath the center plate was etched in the silicon. There are about 40,000 etch holes on this particular device. A close-up image of these etch holes near the edge of the center plate is shown in Figure 8.17.

A large tungsten mass was manually attached to the large center plate of one of the devices fabricated. It is necessary to attach additional mass to the structure to get a useful amount of power out from the target input vibrations. The designs worked out in chapter 6 also assume that additional mass is to be attached. An image of the device with the tungsten block attached is shown in Figure 8.18. The author has not been successful in obtaining test data from this device to demonstrate the functionality of the converter. As with other devices fabricated using this process, a combination of reverse leakage current from the diodes and high resistivity of the structure is thought to be the primary reason that the device has not functioned properly. The device is shown here to demonstrate that the mass can be effectively attached, and to show what a device with a large attached mass looks like.

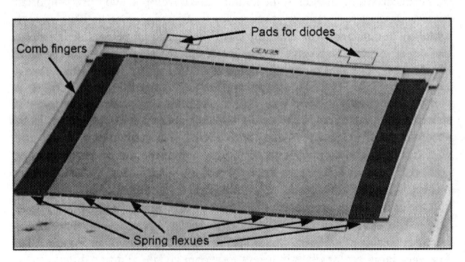

*Figure –8.15*. SEM image of a large (0.25 cm²) electrostatic converter.

*Figure –8.16.* Close-up view of comb fingers and spring flexure.

*Figure –8.17.* Close-up view of etch holes in the center plate on a large converter.

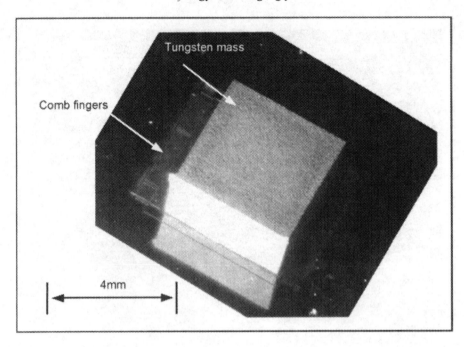

*Figure –8.18.* Image of electrostatic converter with attached tungsten proof mass.

Figure 8.15 shows two pads to which bare die diodes can be attached. As mentioned in chapter 7, two methods of connecting the structure to diodes were used. First, bare die diodes were assembled on the pads shown in Figure 8.15 manually with conductive epoxy, and second the pads shown in Figure 8.15 were used as wire bond pads to connect to off chip surface mount diodes. Figure 8.19 shows a bare die diode attached to an electrostatic converter structure. One of the spring flexures is broken on the structure shown in the image. Both conductive silver epoxy and structural epoxy were used to attach the diode. The epoxy seen around the diode is a structural epoxy applied after the silver conductive epoxy to give the bond added strength for subsequent wire bonding.

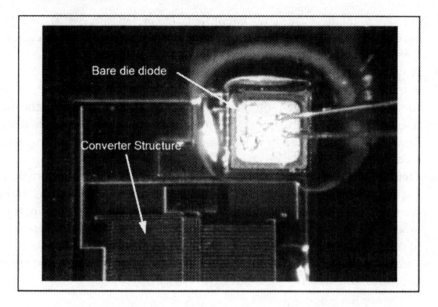

*Figure –8.19.* Electrostatic converter with bare die diodes attached to on-chip pads.

## 5. DISCUSSION OF RESULTS AND CONCLUSIONS

### 5.1 Current Status

The effort to fabricate, micromachined electrostatic vibration-to-electricity converters was not completely successful. A fully functional prototype has yet to be achieved. However, models, design practices, and actual designs have been developed. Electrostatic converter structures have been fabricated on both the meso- and micro-scale. Furthermore, the fundamental voltage (and energy) step-up from mechanical actuation has been demonstrated with both meso- and micro-scale prototypes.

### 5.2 Recommendations

To achieve a fully functional electrostatic converter capable of a useful amount of power output, the following recommendations are made for future designs and processes:

1. Fabricate the diodes on the MEMS chip itself. Integrated diodes are highly preferable because of the drastically reduced parasitic capacitance. The integrated process can be significantly simplified and still achieve

integrated diodes. A new, simplified, integrated process should be developed. Ideally, the diodes will have a lower reverse leakage current than exhibited by the previously fabricated integrated diodes. The process needs to be tuned such that this is the case.

2. It is very important that the silicon be highly doped, and that a metal conductor be deposited on the structure to further reduce parasitic resistances. The new process should include a viable method of depositing metal on the front side as the previously described integrated process does.

3. Etching portions of the substrate away beneath the converter structure has the advantages of reducing the parasitic capacitance and the fluid damping. However, the likelihood of damaging structures during processing is far greater if the substrate is etched away. Therefore, ideally, the process and design could be implemented either with or without a backside etch to remove the substrate beneath converter structure.

As mentioned previously, the principle advantage of electrostatic generators is their potential for implementation in a silicon micromachining process. Implementation in a MEMS process has two advantages: first, there is more potential for monolithic integration with electronics, and second, the cost of production using highly parallel IC fabrication techniques is potentially lower. However, the power production capability of any converter is proportional to the mass of the system. A MEMS implementation, therefore, suffers because of the very low mass of planar devices made of lightweight silicon. It then becomes highly desirable, even necessary, to attach a mass to the MEMS device after fabrication. Even with the attached mass, in order to produce enough power for RF communication from the type of vibrations studied in this work, the area of the entire device needs to be on the order of tens to a hundred square millimeters. However, at this size, it is not cost effective to monolithically integrate the device with electronics. The price of IC's is proportional to the area of silicon they consume. Because the area consumed by electronics would be at least an order of magnitude lower than that consumed by the generator device, the cost of monolithic integration would be inordinately high. Therefore, given the current constraints and application space an electrostatic MEMS implementation of a vibration-to-electricity converter is not economically attractive. However, it is feasible that under a different set of constraints, either a much larger or higher frequency vibration source, or lower power requirements, a MEMS electrostatic converter could become attractive. It is with the understanding of this eventuality that the development of processes and designs for MEMS vibration generators must be pursued.

# Chapter 9

# CONCLUSIONS

This book began with a broad survey of potential power sources for wireless sensor networks. Low level vibrations as a power source were singled out as a power source that merited further research. The remainder of the book, then, has reported on research efforts to model, design, construct, and test vibration to electricity converters. This final chapter will review the justification for this work, its major findings and contributions, and recommendations for future research.

## 1. JUSTIFICATION FOR FOCUS ON VIBRATIONS AS A POWER SOURCE

Power systems represent perhaps the most challenging technological hurdle yet to be overcome in the widespread deployment of wireless sensor networks. While there is still much research to be done improving radio systems for wireless sensor networks, the technology to accomplish a wireless sensor network is currently available. However, even with the aggressive power consumption target of 100 $\mu$W/node, current battery technology cannot even provide 1 year of autonomous operation per 1cm$^3$ of size. Although the energy density of batteries is improving with time, it is doing so very slowly compared to the improvement in size and power consumption of CMOS electronics. Wireless systems have traditionally been designed to use a battery as their power source. However, if ubiquitous wireless sensor networks are to become a reality, clearly alternative power sources need to be employed.

Numerous potential sources for power scavenging exist. Light is routinely used as a power source using photovoltaic cells. Smart cards and RF ID tags, to which power is radiated by an energy rich reader, are also common. Myriads of other sources ranging from thermal gradients to imbalanced AC electric fields may also be imagined. It should be stated clearly that there is no single energy scavenging solution that will provide power in all potential applications. Solutions need to be tailored both to the demands of the application and to the environment in which the system will be used. Based on vibrations measured in many environments, preliminary calculations showed that power densities on the order of 200 $\mu$W/cm$^3$ are feasible from commonly occurring vibrations. This number compares well with other potential energy scavenging sources. For example, in common indoor lighted environments, the power density available from photovoltaic cells is only about 10 $\mu$W/cm$^2$. Likewise, thermoelectric devices can produce about 10 $\mu$W of power from a 10° C temperature differential. Furthermore, vibrations as a power source for stand-alone wireless electronics have received very little research effort. It is believed that the research project reported herein provides an attractive power source for many environments in which low level vibrations are commonly found, and significantly contributes to the development of potentially infinite life power systems for wireless sensor nodes.

Three types of vibration to electricity converters have been considered: electromagnetic, electrostatic, and piezoelectric. After a preliminary investigation, only piezoelectric and electrostatic were pursued in detail. Both types of converters have been modeled, designed, and fabricated. While solar cell based power systems have also been developed for the target wireless sensor nodes, this has been more of a development and benchmarking effort than a research effort.

## 2.    PIEZOELECTRIC VIBRATION TO ELECTRICITY CONVERTERS

Piezoelectric benders that exploit the 31 mode of operation were chosen as a design platform because of the higher strain and lower resonant frequencies that can be generated compared to 33 mode piezoelectric stacks. Based on input vibrations of 2.25 m/s$^2$ at 120 Hz, which represent an average value of the sources measured, a power generation density of 300 $\mu$W/cm$^3$ has been demonstrated. When connected to more realistic power train circuitry, the maximum power transfer to the storage reservoir was 200 $\mu$W/cm$^3$. Furthermore, if more control over the manufacturing process were available (specifically, if piezoelectric layer thickness could be arbitrarily chosen for a given application), simulations of optimal designs demonstrate potential power densities approaching 700 $\mu$W/cm$^3$ from the same vibration source.

For the given application space (i.e. devices on the order of 1cm$^3$, with power generation requirements on the order of 100 μW), piezoelectric converters represent the most attractive solution. Not only are they capable of higher power density, but they are more robust, and the power electronics needed are less complex than electrostatic converters.

## 3. DESIGN CONSIDERATIONS FOR PIEZOELECTRIC CONVERTERS

Detailed models have been developed and validated that can serve as the basis for design and optimization. Practical design relationships result from these models that can serve as guiding principles to the designer. These design principles for piezoelectric converters are summarized below.

1. The power output falls off dramatically if the resonant frequency of the converter does not match that of the driving vibrations. The converter should be designed to resonate at the frequency of the target vibrations.
2. Power output is proportional to the oscillating proof mass. Therefore, the mass should be maximized within the space constraints. The need to design for the frequency of the input vibrations, and to not exceed the yield strain of the piezoelectric material, may also limit the amount of mass that can be used.
3. The power output is inversely proportional to frequency of the driving vibrations. Therefore, the system should be designed to resonate at the lowest frequency peak in the input vibration spectrum provided that higher frequency peaks do not have a larger acceleration magnitude.
4. The energy removed from the oscillating converter can be treated as electrically induce damping. Optimal power transfer occurs when the effective electrically induced damping ratio is equal to the mechanical damping ratio. In the case of a simple resistive load, the electrically induced damping is a function of the load resistance, and so can be set by properly choosing the load resistance. In other cases, different circuit parameters can be changed which will affect the amount of electrically induced damping.
5. A storage capacitor charged up through a full wave rectifier is a fairly realistic load. Optimal power transfer to the storage capacitor occurs when the voltage across it is approximately one half the open circuit voltage of the piezoelectric converter. The load circuitry should be designed to control, or at least set limits on, the voltage range of the storage capacitor.

6. The size of the storage capacitor should be at least 10 times the capacitance of the piezoelectric device. If it is smaller than about 10 times the capacitance of the piezoelectric device, the power transfer to the storage capacitor increases with increasing capacitance. If it is greater than about 10 times the capacitance of the converter, the power transfer is largely unaffected by the size of the storage capacitor. A good rule of thumb is to design the storage capacitor to be at least 100 times the capacitance of the piezoelectric device. Additionally, the storage capacitor needs to be large enough to supply enough charge to the load during "on" cycles without its voltage dropping too far. This second consideration will probably drive the storage capacitor to be far greater than 100 times the capacitance of the piezoelectric converter.

## 4.    ELECTROSTATIC VIBRATION TO ELECTRICITY CONVERTERS

The design of electrostatic converters was pursued primarily because they are easily implemented in silicon micromachining technology. The utilization of this fabrication technology offers the significant potential benefit of future monolithic integration with sensors and electronics. However, the fundamental conversion potential for electrostatic converters is lower than for piezoelectric converters. As explained in chapter 3, the maximum potential energy density for electrostatic transducers is lower than for piezoelectric converters. This fact is further demonstrated by the simulated power output of optimal electrostatic and piezoelectric designs. Simulations show that optimized electrostatic designs can produce about 110 $\mu$W/cm$^3$ while optimized piezoelectric designs can potentially produce several times that value. Nevertheless, models have been developed, designs performed, and devices fabricated. Devices have been fabricated using a SOI MEMS technique, and the basic operational principles of the converters have been demonstrated. However, power output on the same order of magnitude as that predicted by models has not yet been achieved.

In order for an electrostatic implementation to become attractive, some combination of the following would need to occur. First, in order for monolithic integration to be economically feasible, the size of the converter would need to be reduced to around 1 mm$^3$ rather than 1 cm$^3$. Therefore, the potential power production would decrease by a factor of 1000, or to about 100 nW. If stand-alone systems that can function effectively on 100 nW of power can be implemented, then electrostatic generators could be attractive. Designing the converters to resonate at around 120 Hz in 1mm$^3$ could however, be a very significant challenge. Second, in certain environments vibrations of far higher acceleration amplitudes and frequencies are available. If the vibration sources are more energetic, the same amount of power could be generated with a smaller device. If enough applications with high level vibrations arise, electrostatic generators may be the preferred implementation. At present, it is not clear that either of these two situations will exist or that electrostatic converters will become an equally attractive alternative to piezoelectric converters.

## 5.    DESIGN CONSIDERATIONS FOR ELECTROSTATIC CONVERTERS

As with piezoelectric converters, detailed models have been developed for electrostatic in-plane gap closing generators. Based on these models, and the general model developed in Chapter 2, the following design guidelines emerge.

As in the case of all resonant vibration-to-electricity conversion, an electrostatic converter should be designed to resonate at the frequency of the driving vibrations. While, this consideration is perhaps less critical than it is in the case of piezoelectric converters because of the lower effective quality factor of in-plane gap closing electrostatic converters, it is nevertheless important.

Again, power output is proportional to mass. This consideration presents a more challenging problem in the domain of micromachined electrostatic converters because of the planar nature of most MEMS devices and the low density of silicon. Therefore, a means of attaching additional mass to the device needs to be employed.

The power output is related to the ratio of maximum to minimum capacitance of the converter device. Therefore, it is generally beneficial to design the converter device such that the maximum capacitance is as high as possible. As parasitic capacitance can easily swamp the device making the voltage gain across the variable capacitor negligible, it is also generally beneficial to minimize the parasitic capacitance associated with the device.

The range of motion of the device is determined by the nominal gap between comb fingers, the placement of mechanical limit stops, and the mechanical dynamics of the system. The level of effective electrically induced damping is related to the range of motion of the device, and therefore an optimal nominal gap between comb fingers exists. There is strong interaction between the device geometry (such as device thickness and finger length), the mechanical dynamics, and the choice of the nominal gap. Therefore, dynamic simulations and optimization routines need to be used to properly choose the nominal gap.

## 6.      SUMMARY OF CONCLUSIONS

The above conclusions may be thus summarized:

1. Virtually all of the vibration sources measured in conjunction with this study have a dominant frequency in the range of 70 Hz to 125 Hz and magnitudes on the order of tenths to several $m/s^2$.
2. Power densities on the order of 200 $\mu W/cm^3$ are possible and have been demonstrated using piezoelectric converters from an input vibration source of 2.25 $m/s^2$ at 120 Hz.
3. Given the current set of design constraints, which are thought to represent most potential wireless sensor node applications, piezoelectric converters are the preferred technology because of their higher power density and simpler power electronics.


## 7.      RECOMMENDATIONS FOR FUTURE WORK

Energy scavenging represents an ideal power solution for wireless sensor networks because of its potential to power the sensor nodes indefinitely. While the field of energy scavenging is much more broad than vibration to electricity conversion, it is the authors' opinion that there is a wide variety of applications that could greatly benefit from vibration based generators. With this in mind, much more work can be done to advance the field. A few issues that have yet to be satisfactorily pursued are presented below.

As suitable applications currently exist for piezoelectric converters, it is the authors' opinion that immediate research and development work should focus on piezoelectric converters until it can be shown that conditions will likely exist that would make electrostatic converters preferable. Only a very limited number of design configurations have been evaluated in this study. Other design configurations that are have better fatigue characteristics, and potentially higher power outputs should be evaluated. Although it is more difficult to fabricate piezoelectric converters on a silicon chip with a micromachining process, it is by no means impossible. It is felt that an effort to both improve thinfilm PZT (and other piezoelectric materials) processes and redesign the converter for a microfabrication process using PZT would be more justified than focusing effort on the development of an electrostatic converter.

Converters have been designed to resonate at the frequency of the driving vibrations. However, this frequency must be known for this approach to work. It is, of course, possible to actively tune the resonant frequency, however, research on space and energy efficient methods to accomplish this needs to be undertaken. Additionally, extremely low power control techniques to accomplish this tuning also need to be investigated. An open question is whether or not the net power output will be greater with active frequency tuning.

Finally, most commercial power electronics components (i.e. voltage regulators and DC-DC converters) are optimized for systems with much higher power dissipation. Indeed, most standard designs in textbooks are not optimized for systems with average power consumptions well below 1mW. Power electronics specifically optimized for vibration to electricity conversion based on piezoelectrics could significantly improve the overall performance of the system. Furthermore, the system could significantly benefit from an extremely large inductive structure in series with the piezoelectric converter. Given the low frequency of oscillation, the size of the inductor would be too large to use traditional inductors. However, micro-mechanical structures with very large effective inductances can be fabricated and could, perhaps, improve the power transfer to the load if placed in series with the piezoelectric converter.

In a recent article in IEEE Computer, Professor Jan Rabaey states "One of the most compelling challenges of the next decade is the 'last meter' problem—extending the expanding data network into end-user data-collection and monitoring devices." (Rabaey *et al* 2000). Indeed, it has become a widely held belief in the research and business community that the next revolution in computing technology will be the widespread deployment of low cost, low power wireless computing devices. In order for this vision to become a reality, the problem of how to power the devices in a cost effective way must be solved. Primary batteries are appropriate for a certain class of devices. However, if the wireless sensing and computing nodes are to become truly ubiquitous, the replacement of batteries is simply too costly. Energy scavenging technologies must be developed in order to create completely self-sustaining wireless sensor nodes. The research presented in this book has demonstrated that low level vibrations can provide enough power to operate wireless sensor nodes in many applications. However, as stated previously, there is no single energy scavenging solution that will fit all applications and all environments. Energy scavenging solutions must, therefore, continue to be explored in order to meet the needs of an ever growing application space for wireless sensor networks.

# Acknowledgments

This work could never have come to fruition without the generous support of a great many people, whose efforts are greatly appreciated. We would specifically like to acknowledge the contributions of the following people and organizations.

Professors Kris Pister and Seth Sanders have been generous with their time, have provided many important insights, and have aided in the direction of the research.

Brian Otis developed the RF transceiver that has been integrated with the vibration based generator as reported in chapter 5. Additionally he has generously assisted with our understanding of circuit design issues and instrumentation.

Fred Burghardt and Sue Mellers have been very helpful in the implementation of power and measurement circuits.

Elizabeth Reilly, Eric Mellers, and Sid Mal have been instrumental in the implementation of converter prototypes and system integration.

Professor Luc Frechette and Dan Stengart have contributed to our understanding of benchmarking technologies, specifically micro-engines, micro-batteries, and micro-fuel cells.

The authors gratefully acknowledge the Intel Corporation for support under the Noyce Fellowship, the US Department of Energy for support under the Integrated Manufacturing Fellowship, and the support of DARPA under grant # F33615-02-2-4005.

# Appendix A: Analytical Model of a Piezoelectric Generator

Chapter 4 discussed the development of an analytical model for piezoelectric generators and used this model as a basis for design. However, many of the details of the derivation of the analytical model were left out of chapter 4 to improve the readability of the chapter. The goal of this appendix, then, is to provide the details of the derivation of the analytical model for a piezoelectric vibration-to-electricity converter.

## 1. GEOMETRIC TERMS FOR BIMORPH MOUNTED AS A CANTILEVER

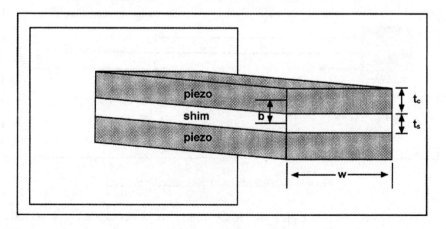

*Figure –A.1.* Cross section of composite beam.

Because the piezoelectric bender is a composite beam, an effective moment of inertia and elastic modulus are used. The effective moment of inertia is given by equation A.1 below.

$$I = 2\left[\frac{wt_c^3}{12} + wt_c b^2\right] + \frac{\eta_s wt_{sh}^3}{12} \qquad (A.1)$$

where $w$ is the width of the beam, $t_c$ is the thickness of an individual piezoelectric ceramic layer, $b$ is the distance from the center of the shim to the center of the piezo layers, $t_{sh}$ is the thickness of the center shim, and $\eta_s$ is the ratio of the piezoelectric material elastic constant to that of the center shim ($\eta_s = Y_c/Y_{sh}$ where $Y_c$ is Young's modulus for the piezoelectric ceramic and $Y_{sh}$ is Young's modulus for the center shim).

The elastic constant for the piezoelectric ceramic is then used in conjunction with the effective moment of inertia shown by equation A.1. The different Young's modulus of the center shim is accounted for by the term $\eta_s$ in the moment of inertia (Beer and Johnston 1992).

Because the piezoelectric constitutive equations deal directly with stress and strain, it is most convenient to use them as the state equations for the dynamic system rather than force and displacement. However, in order to derive state equations in terms of stress and strain for the piezoelectric bender mounted as a cantilever beam as shown in Figure A.2, two geometric terms need to be defined. The first relates vertical force to average stress in the piezoelectric material, and the second relates tip deflection of the beam to average strain in the piezoelectric material.

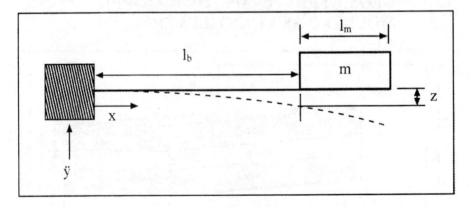

*Figure –A.2.* Schematic of piezoelectric bender.

When purchased, the piezoelectric benders are covered with an electrode. The electrode can be easily etched away making the only the portion of the beam covered by the electrode active as a piezoelectric element. Referring to Figure A.2, it will be assumed that the electrode length ($l_e$, not shown in the figure) is always equal to or less than the length of the beam ($l_b$ in the figure). The stress and strain values of interest, and those used as state variables, are the average stress and strain in the piezoelectric material that is covered by an electrode. An expression for the average stress in the piezoelectric material covered by the electrode (hereafter referred to simply as stress) is given by equation A.2.

$$\sigma = \frac{1}{l_e} \int_0^{l_e} \frac{M(x)b}{I} dx \tag{A.2}$$

where $\sigma$ is stress, $x$ is the distance from the base of the beam, and $M(x)$ is the moment in the beam as a function of the distance ($x$) from its base.

The moment, $M(x)$, is given by equation A.3.

$$M(x) = m(\ddot{y} + \ddot{z})(l_b + \frac{1}{2}l_m - x) \tag{A.3}$$

where $l_m$ is the length of the mass, $\ddot{y}$ is the input vibration in terms of acceleration, and $z$ is the vertical displacement of the beam at the point where the mass attaches with respect to the base of the beam.

Substituting equation A.3 in to equation A.2 yields the expression in A.4.

$$\sigma = m(\ddot{y} + \ddot{z}) \frac{b(2l_b + l_m - l_e)}{2I} \tag{A.4}$$

The vertical force term in equation A.4 is simply $m(\ddot{y} + \ddot{z})$. Therefore, let us define $b^{**}$ as shown in equation A.5.

$$b^{**} = \frac{2I}{b(2l_b + l_m - l_e)} \tag{A.5}$$

$b^{**}$ then relates vertical force to stress ($\sigma$), and $\sigma = m(\ddot{y} + \ddot{z})/b^{**}$.

In order to derive the term relating deflection at the point where the beam meets the mass as shown in Figure A.2 to average strain in the piezoelectric material covered by the electrode (simply strain hereafter), consider the Euler beam equation shown as A.6.

$$\frac{d^2z}{dx^2} = \frac{M(x)}{Y_cI} \tag{A.6}$$

Substituting A.3 into A.6 yields equation A.7.

$$\frac{d^2z}{dx^2} = \frac{1}{Y_cI}m(\ddot{y}+\ddot{z})(l_b + \frac{1}{2}l_m - x) \tag{A.7}$$

Integrating to obtain an expression for the deflection term ($z$) yields:

$$z = \frac{m(\ddot{y}+\ddot{z})}{Y_cI}\left((l_b + \frac{1}{2}l_m)\frac{x^2}{2} - \frac{x^3}{6}\right) \tag{A.8}$$

At the point where the beam meets the mass (at $x = l_b$), the expression for $z$ becomes:

$$z = \frac{m(\ddot{y}+\ddot{z})l_b^2}{2Y_cI}\left(\frac{2}{3}l_b + \frac{1}{2}l_m\right) \tag{A.9}$$

Finally, realizing that strain is equal to stress over the elastic constant, $\delta = \sigma/Y$, and that stress can be expressed as in equation A.4, strain can be written as shown below:

$$\delta = \frac{m(\ddot{y}+\ddot{z})b}{2Y_cI}(2l_b + l_m - l_e) \tag{A.10}$$

Rearranging equation A.10, the force term, $m(\ddot{y}+\ddot{z})$, can be written as shown in equation A.11.

$$m(\ddot{y}+\ddot{z}) = \frac{2Y_cI}{b(2l_b + l_m - l_e)}\delta \tag{A.11}$$

Substituting equation A.11 into equation A.9 yields:

$$z = \delta\frac{l_b^2}{3b}\frac{(2l_b + \frac{3}{2}lm)}{(2l_b + l_m - l_e)} \tag{A.12}$$

Let us define $b^*$ as shown in equation A.13. $b^*$ then relates strain to vertical displacement, or $z = \delta/b^*$.

$$b^* = \frac{3b}{l_b^2} \frac{(2l_b + l_m - l_c)}{(2l_b + \frac{3}{2}l_m)} \tag{A.13}$$

## 2. BASIC DYNAMIC MODEL OF PIEZOELECTRIC GENERATOR

A convenient method of modeling piezoelectric elements such that system equations can be easily developed is to model both the mechanical and electrical portions of the piezoelectric system as circuit elements. The piezoelectric coupling is then modeled as a transformer (Flynn and Sanders 2002). Figure A.3 shows the circuit model for a piezoelectric element. Note that this is the same circuit model shown in chapter 4 as Figure A.3. Note also that no electric load is applied to the system.

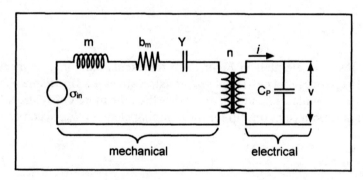

*Figure –A.3.* Circuit representation of piezoelectric bimorph

As explained in chapter 4, the across variable (variable acting across an element) on the electrical side is voltage ($V$) and the through variable (variable acting through an element) is current ($i$) (Rosenberg and Karnopp 1983). The across variable on the mechanical side is stress ($\sigma$) and the through variable is strain ($\delta$). The system equations can then be obtained by simply applying Kirchoff's Voltage Law (KVL) and Kirchoff's Current Law (KCL). However, first the stress / strain relationships for circuit elements on the mechanical side need to be defined.

$\sigma_{in}$ is an equivalent input stress. In other words, it is the stress developed as a result of the input vibrations. $m$, shown as an inductor, represents the effect of the mass, or inertial term. The stress "across" this element is the stress developed as a result of the mass flexing the beam. Equation A.4 gives the stress resulting from both the input element, $\sigma_{in}$, and the inertial element, $m$. Thus, the relationships for these two elements are given in equations A.14 and A.15.

$$\sigma_{in} = \frac{m}{b^{**}} \ddot{y} \qquad (A.14)$$

$$\sigma_m = \frac{m}{b^{**}} \ddot{z} \qquad (A.15)$$

However, the preferred state variable is strain, $\delta$, rather then displacement $z$. Substituting strain for displacement in A.15 using the relationship from equation A.12 and A.13 yields the stress / strain relationship for the inertial element, m, in equation A.16.

$$\sigma_m = \frac{m}{b^* b^{**}} \ddot{\delta} \qquad (A.16)$$

The resistive element in Figure A.3 represents damping, or mechanical loss. The damping coefficient, $b_m$, relates stress to tip displacement, $z$. Therefore the units of $b_m$ are Ns/m$^3$ rather than the more conventional Ns/m. The stress / strain relationship for the damping element, $b_m$, becomes:

$$\sigma_{bm} = \frac{b_m}{b^*} \dot{\delta} \qquad (A.17)$$

Finally, the stiffness element is represented as a capacitor and labeled with the elastic constant, $Y$. The stress / strain relationship for this element is simply Hooke's law, shown here as equation A.18.

$$\sigma_Y = Y_c \delta \qquad (A.18)$$

The transformer relates stress ($\sigma$) to electric field ($E$) at zero strain, or electrical displacement ($D$) to strain ($\delta$) at zero electric field. The piezoelectric constitutive relationships are shown again below in equations A.19 and A.20. The equations for the transformer follow directly from the piezoelectric constitutive relationships and are given in equations A.21 and A.22.

$$\delta = {}^{\sigma}\!/_{Y} + dE \tag{A.19}$$

$$D = \varepsilon E + d\sigma \tag{A.20}$$

where, $d$ is the piezoelectric strain coefficient, and $\varepsilon$ is the dielectric constant of the piezoelectric material.

$$\sigma_t = -dY_c E \tag{A.21}$$

$$D_t = -dY_c \delta \tag{A.22}$$

The equivalent turns ratio for the transformer is then $-dY$. The state variables, however, are current, $\dot{q}$, and voltage, $V$. Noting that $q = l_e w D$ (or $q = 2l_e w D$ if the layers are poled in parallel), and that $V = 2Et_c$ (or $V = Et_c$ if wired in parallel) the equations for the transformer can be rewritten as:

$$\sigma_t = \frac{-adY_c}{2t_c} V \tag{A.23}$$

$$\dot{q}_t = -dY_c a l_e w \dot{\delta} \tag{A.24}$$

where $a = 1$ if the two piezoelectric layers are wired in series, and $a = 2$ if they are wired in parallel.

Applying KVL to the circuit in Figure A.3 yields the following equation:

$$\sigma_{in} = \sigma_m + \sigma_{bm} + \sigma_t \tag{A.25}$$

Substituting equations A.14, A.16 – A.18, and A.23 into A.25 and rearranging terms yields the third order equation shown as A.26, which describes the mechanical dynamics of the system with an electrical coupling term.

$$\ddot{\delta} = -\frac{Yb^{*}b^{**}}{m}\delta - \frac{b_m b^{**}}{m}\dot{\delta} + \frac{daY_c}{2t_c}\frac{b^{*}b^{**}}{m}V + b^{*}\ddot{y} \qquad (A.26)$$

The combined term $Yb^{*}b^{**}$ has units of force / displacement and relates vertical force to tip deflection. This is commonly referred to as the effective spring constant. Letting $k_{sp}$ be the effective spring constant, and substituting $k_{sp} = Yb^{*}b^{**}$ into equation A.26 yields the simpler and more familiar expression in equation A.27.

$$\ddot{\delta} = -\frac{k_{sp}}{m}\delta - \frac{b_m b^{**}}{m}\dot{\delta} + \frac{k_{sp}da}{2mt_c}V + b^{*}\ddot{y} \qquad (A.27)$$

Equation A.27 forms a portion of the complete dynamic model. Applying KCL to the electrical side of the circuit in Figure A.3 yields the rest of the model. Equation A.28 is the very simple result of applying KCL to the electrical side of the equivalent circuit.

$$\dot{q}_t = \dot{q}_{Cp} \qquad (A.28)$$

where $\dot{q}_t$ is the current through the transformer as defined in equation A.24, and $\dot{q}_{Cp}$ is the current through the capacitor $C_p$.

The capacitance of the piezoelectric device is $C_p = a^2 \varepsilon w l_e / 2t_c$. Substituting equation A.24 into A.28, using the long expression for $C_p$, and rearranging terms yields the equation shown as A.29.

$$\dot{V} = \frac{-2dY_c t_c}{a\varepsilon}\dot{\delta} \qquad (A.29)$$

Equations A.27 and A.29 constitute the dynamic model of the system. They can be rewritten in state space form as shown in equation A.30.

$$\begin{bmatrix} \dot{\delta} \\ \ddot{\delta} \\ \dot{V} \end{bmatrix} = \begin{bmatrix} 0 & 1 & 0 \\ -\dfrac{k_{sp}}{m} & -\dfrac{b_m b^{**}}{m} & \dfrac{k_{sp}da}{2mt_c} \\ 0 & -\dfrac{2dY_c t_c}{a\varepsilon} & 0 \end{bmatrix} \begin{bmatrix} \delta \\ \dot{\delta} \\ V \end{bmatrix} + \begin{bmatrix} 0 \\ b^{*} \\ 0 \end{bmatrix} \ddot{y} \qquad (A.30)$$

As noted previously, no electrical load has been applied to the system. The right side of Figure A.3 is an open circuit, and so no power is actually transferred in this case. Figure A.4 shows the circuit representation of the system with a simple resistive load applied.

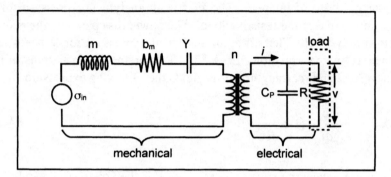

*Figure –A.4.* Circuit model for piezoelectric bimorph with resistive load.

The resulting change in the system equations is only minor. The mechanical side of the circuit, and thus equation A.27, remain unchanged. Applying KCL to the electrical side of the circuit now yields:

$$\dot{q}_i = \dot{q}_{Cp} + \dot{q}_R \tag{A.31}$$

where, $\dot{q}_R$ is the current through resistor $R$.

Making the same substitutions as explained previously, and substituting $V/R$ for the current through the resistor yields the following equation, which replaces equation A.29 in the system model.

$$\dot{V} = \frac{-2Y_c dt_c}{a\varepsilon}\dot{\delta} - \frac{1}{RC_p}V \tag{A.32}$$

The new system model in state space form is given by equation A.33.

$$\begin{bmatrix} \dot{\delta} \\ \ddot{\delta} \\ \dot{V} \end{bmatrix} = \begin{bmatrix} 0 & 1 & 0 \\ -\dfrac{k_{sp}}{m} & -\dfrac{b_m b^{**}}{m} & \dfrac{k_{sp} da}{2mt_c} \\ 0 & -\dfrac{2dY_c t_c}{a\varepsilon} & \dfrac{-1}{RC_p} \end{bmatrix} \begin{bmatrix} \delta \\ \dot{\delta} \\ V \end{bmatrix} + \begin{bmatrix} 0 \\ b^* \\ 0 \end{bmatrix} \ddot{y} \tag{A.33}$$

## 3.     EXPRESSIONS OF INTEREST FROM BASIC DYNAMIC MODEL

A few additional expressions derived from the basic model shown in equation A.33 are of interest. The first is an analytical expression for the power transferred to the resistive load. The power dissipated by the resistive load is simply $V^2/R$. Therefore, an analytical expression for V needs to be obtained from the equations in A.33. Taking the Laplace transform of equation A.32 and rearranging terms yields the following expression:

$$\Delta = -\frac{a\varepsilon}{2Y_c dt_c s}(s + \frac{1}{RC_p})V \tag{A.34}$$

where, $\Delta$ is Laplace transform of strain ($\delta$), V is the voltage (the symbol V is used in both the time and frequency domain), and s is the Laplace variable.

Taking the Laplace transform of equation A.26 and rearranging terms yields:

$$\Delta\left(s^2 + \frac{b_m b^{**}}{m}s + \frac{k_{sp}}{m}\right) = \frac{k_{sp}da}{2mt_c}V + b^* A_{in} \tag{A.35}$$

where $A_{in}$ is the Laplace transform of the input vibrations in terms of acceleration.

Substituting equation A.34 into A.35 and rearranging terms yields the following expression:

$$V\left[s^3 + \left(\frac{1}{RC_p} + \frac{b_m b^{**}}{m}\right)s^2 + \left(\frac{k_{sp}}{m}\left(1 + \frac{d^2 Y_c}{\varepsilon}\right) + \frac{b_m b^{**}}{mRC_p}\right)s + \frac{k_{xp}}{mRC_p}\right] = -\frac{2Y_c dt_c b^*}{a\varepsilon}A_{in} \tag{A.36}$$

The expression in equation A.36 can be solved for the output voltage. The resulting expression is perhaps more meaningful with the following substitutions: $d^2 Y_c/\varepsilon$ is the square of a term commonly referred to as the piezoelectric coupling coefficient denoted by the symbol $k$, the Laplace variable may be substituted with $j\omega$ where $j$ is the imaginary number, $k_{sp}/m$ is the natural frequency of the system represented by the symbol $\omega_n$, and the damping term $b_m b^{**}/m$ can be rewritten in terms of the unitless damping ratio $\zeta$ as $2\zeta\omega_n$. Making these substitutions and solving for V yields:

$$V = \cfrac{-j\omega\cfrac{2Y_c dt_c b^*}{a\varepsilon}}{\left[\cfrac{1}{RC_p}\omega_n^2 - \left(\cfrac{1}{RC_p} + 2\zeta\omega_n\right)\omega^2\right] + j\omega\left[\omega_n^2(1+k^2) + \cfrac{2\zeta\omega_n}{RC_p} - \omega^2\right]}A_{in}$$

(A.37)

If the further simplifying assumption is made that the resonant frequency $\omega_n$ matches the driving frequency $\omega$, equation A.37 reduces to:

$$V = \cfrac{-j\cfrac{2Y_c dt_c b^*}{a\varepsilon}}{2\zeta\omega^2 + j\left[\omega^2 k^2 + \cfrac{2\zeta\omega}{RC_p}\right]}A_{in}$$

(A.38)

As mentioned earlier, the power transferred is simply $V^2/R$. Therefore, using the expression in equation A.38, the resulting analytical term for the magnitude of the power transferred to the load is as follows.

$$P = \cfrac{1}{\omega^2}\cfrac{RC_p^2\left(\cfrac{2Y_c dt_c b^*}{a\varepsilon}\right)^2}{(4\zeta^2 + k^4)(RC_p\omega)^2 + 4\zeta k^2(RC_p\omega) + 2\zeta^2}A_{in}^2$$

(A.39)

The optimal load resistance can then be found by differentiating equation A.38 with respect to $R$, setting the result equal to zero, and solving for $R$. The resulting optimal load resistance is shown in equation A.40.

$$R_{opt} = \cfrac{1}{\omega C_p}\cfrac{2\zeta}{\sqrt{4\zeta^2 + k^4}}$$

(A.40)

An equivalent electrically induced damping ratio, $\zeta_e$, can also be derived from the analytical system model. The electrical coupling term, $k_{sp}daV/2mt_c$, in equation A.27 can be used to find the equivalent linear damping ratio. It written in terms of an electrically induced damping ratio, $\zeta_e$, the electrical coupling term would have the following standard form:

$$\frac{k_{sp}da}{2mt_c}V = 2\zeta_e\omega\dot{\delta} \tag{A.41}$$

The damping ratio, $\zeta_e$, can then be expressed as:

$$\zeta_e = \frac{k_{sp}da}{2mt_c\omega}\frac{V}{\dot{\delta}} \tag{A.42}$$

An expression, in terms of the Laplace variable $s$, for the voltage to strain rate ratio follows from equation A.34, and is shown below.

$$\frac{V}{\Delta s} = -\frac{2Y_cdt_c}{a\varepsilon(s+\dfrac{1}{RC_p})} \tag{A.43}$$

'Note that the term $\Delta s$ in equation A.43 is the same term as $\dot{\delta}$ in equation A.42. Substituting A.43 into A.42, and making the substitution $k^2 = d^2Y_c/\varepsilon$ as before, the resulting expression for the electrically induced damping ratio is:

$$\zeta_e = \frac{\omega k^2}{2\left(s+\dfrac{1}{RC_p}\right)} \tag{A.44}$$

As seen from equation A.44, the damping ratio is a function of frequency and is not necessarily real. The magnitude of damping ratio is shown below in equation A.45, which is the same as shown in chapter 4 as equation 4.11. It should finally be noted that if the expression for the optimal load resistance shown in equation A.40 is substituted into equation A.44, the result is that the magnitude of the electrically induced damping ratio, $\zeta_e$, in fact is equal to the mechanical damping ratio, $\zeta$, as expected.

$$\zeta_e = \frac{\omega k^2}{2\sqrt{\omega^2 + \dfrac{1}{(RC_p)^2}}} \tag{A.45}$$

# 4. ALTERATIONS TO THE BASIC DYNAMIC MODEL

Two significant alterations to the basic dynamic model are of interest. As discussed in chapter 4, it is useful to build a dynamic model with a rectifier and storage capacitor as a load rather than a simple resistor.

The changes to the basic dynamic system equations that result from a rectifier and capacitive load are discussed in sufficient detail in chapter 4 for the reader to be able to duplicate the derivation. The new system equations are given in chapter 4 as equations 4.15 through 4.17. Furthermore, analytical expressions for the charge and energy transferred to the storage capacitor are derived in significant detail in chapter 4 and are shown as equations 4.19 and 4.20. However, the jump from the energy transferred per half cycle in equation 4.20 to the power conversion in equation 4.21 is significant and more detail needs to be provided here.

First, the expressions for charge and energy transferred are in terms of $V_1$ and $V_2$, which represent the voltage across the storage capacitor at the beginning and end of the half cycle respectively. However, because the value of $V_2$ is an unknown at the beginning of the half cycle, it must be calculated or substitutions must be made to remove $V_2$ from the analytical expressions. Reconsider the expression for charge transferred per half cycle given as equation 4.19 and repeated here as equation A.46.

$$\Delta Q = \int_{t_1}^{t_2} i\, dt = \int_{t_1}^{t_2} C_p \frac{d(V_s - V)}{dt} dt \qquad (A.46)$$

This expression can be simplified and rewritten as shown in equation A.47.

$$\Delta Q = C_p \int_{t_1}^{t_2} dV_s - C_p \int_{t_1}^{t_2} dV \qquad (A.47)$$

Remembering that the effective source voltage, $V_s$, as shown in Figure 4.16 of chapter 4 and defined by equation 4.18, can be expressed as $V_s sin(\omega t)$, equation A.47 can be integrated and rewritten as:

$$\Delta Q = C_p \left( V_s \sin(\omega t_2) - V_s \sin(\omega t_1) - V(t_2) + V(t_1) \right) \qquad (A.48)$$

As in chapter 4, $V(t_2)$ and $V(t_1)$ will hereafter be referred to as $V_2$ and $V_1$ for simplicity. $t_2$ is the time at which the rectification diodes turn off at the end of the half cycle. This will always occur at the top of the sinusoid that defines the effective source voltage. In other words $V_s sin(\omega t_2) = V_s$ for any half cycle. Also, as stated in chapter 4, $\Delta Q$ is also equal to $C_{st}(V_2 - V_1)$. $t_1$ is the time at which the effective source voltage, $V_s$, is equal to the voltage across the storage capacitor, $V$. In other words, the time at which the rectification diodes begin to conduct (assuming ideal diodes). At any $t_1$ (that is the time at which $V_s = V$ for any half cycle), the voltage across the storage capacitor, $V_1$, is equal to $V_s sin(\omega t_1)$. Substituting these three equivalencies into equation A.48 yields the expression in equation A.49.

$$V_2 - V_1 = \frac{C_p}{C_{st}}(V_s - V_1 - V_2 + V_1) \tag{A.49}$$

Equation A.49 can now be solved for $V_2$, which results in the following expression:

$$V_2 = \frac{C_{st}V_1 + C_p V_s}{C_{st} + C_p} \tag{A.50}$$

This expression can be used in a simple calculation to generate a voltage versus time curve. Initially $V_1$ is zero, $C_{st}$ and $C_p$ are constants, and $V_s$ is predetermined by the magnitude of the input vibrations and the beam equations. $V_2$ can then be calculated for the half cycle. $V_2$ then becomes $V_1$ for the next half cycle.

The analytical expression for power transferred can now be derived. The expression for energy per half cycle developed in chapter 4 is rewritten here as equation A.51.

$$\Delta E = \frac{1}{2}C_{st}(V_2^2 - V_1^2) \tag{A.51}$$

Recall that the expression for power transferred is $P = 2f\Delta E$. Equation A.50 can be substituted into A.51 and the resulting expression for energy can be placed into the above equation for power. Rearranging terms, the resulting expression for power is shown in equation A.52, which is the same expression shown in chapter 4 as equation 4.21.

$$P = \frac{\omega C_{st}}{2\pi (C_{st} + C_p)^2} \left\{ C_p^2 V_s^2 + 2C_{st}C_p V_s V_1 - C_p V_1^2 (2C_{st} + C_p) \right\} \tag{A.52}$$

The second alteration to the basic model of interest adjusts for the compliance of the cantilever mounting. For smaller generators, a pin-pin mounting model more accurately models the physical system. A schematic of this mounting is shown in Figure A.5.

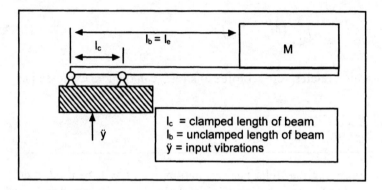

*Figure –A.5.* Illustration of pin-pin mounting model for a piezoelectric generator.

Dimensions and variables not shown in Figure A.5 are the same as shown in Figures 1 and 2. This adjusted mounting model does not affect the circuit representations shown in Figures A.3 and A.4, it only affects the geometric constants relating vertical force to average stress ($b^{**}$) and tip deflection to average strain ($b^{*}$). Therefore, only these two terms need to be re-derived. The dynamic models, as shown in equations A.30 and A.33, are unchanged, but the expressions defining $b^{*}$ and $b^{**}$, and thus $k_{sp}$, are different.

In deriving an expression for the average stress in the piezoelectric material covered by the electrode, it will be assumed the length of the electrode, $l_e$, is equal to the total length of the beam, $l_b$. The expression for the stress is then given by equation A.53.

$$\sigma = \frac{1}{l_c} \int_0^{l_c} \frac{M_1(x)b}{I} \, dx + \frac{1}{l_e - l_c} \int_{l_c}^{l_e} \frac{M_2(x)b}{I} \, dx \tag{A.53}$$

where $M_1(x)$ is the moment function between the two pin mounts (0 and $l_c$), and $M_2(x)$ is the moment function between the right pin and the mass ($l_c$ to $l_b$).

The functions for $M_1(x)$ and $M_2(x)$ are given in equations A.54 and A.55.

$$M_1(x) = m(\ddot{y} + \ddot{z}) \left( \frac{2l_b + l_m}{2l_c} \right) x \tag{A.54}$$

$$M_2(x) = m(\ddot{y} + \ddot{z})\left(x - a - l_b - \frac{l_m}{2}\right) \tag{A.55}$$

Substituting equations A.54 and A.55 into equation A.53 yields:

$$\sigma = m(\ddot{y} + \ddot{z})\frac{b}{4I}\left(4l_b + 3l_m\right) \tag{A.56}$$

$b^{**}$, which relates vertical force to stress, can then be expressed as:

$$b^{**} = \frac{4I}{b\left(4l_b + 3l_m\right)} \tag{A.57}$$

In considering the Euler beam equation, which will be used to derive the expression for $b^{*}$, let $z_1$ be the vertical deflection from $x = 0$ to $l_c$, and $z_2$ be the vertical deflection from $x = l_c$ to $l_b$. $M_1(x)$ then corresponds to $z_1$, and $M_2(x)$ corresponds to $z_2$. The Euler beam equation, as shown above in equation A.6 can be applied to each section of the beam. Substituting equations A.54 and A.55 into the two Euler beam equations, and integrating yields the expression in equation A.58 for the vertical deflection of the beam between $x = l_c$ and $x = l_b$. It should be noted that one of the boundary conditions for $z_2$ is that $dz_2/dx\ (x=l_c) = dz_1/dx\ (x=l_c)$.

$$z_2 = \frac{m(\ddot{y} + \ddot{z})}{Y_c I}\left[x^2\left(\frac{x}{6} - \frac{l_c}{2} - \frac{l_b}{2} - \frac{l_m}{4}\right) + l_c x\left(\frac{l_c}{2} + \frac{2l_b}{3} + \frac{l_m}{3}\right)\right.$$
$$\left. - \frac{l_c^2}{12}\left(2l_c + 2l_b + l_m\right)\right] \tag{A.58}$$

At the point where the mass attaches to the beam, at $x = l_b$, the expression for vertical deflection reduces to:

$$z = \frac{m(\ddot{y} + \ddot{z})l_b}{12Y_c I}\left(4l_b + 3l_m\right)\left(l_c + l_b\right) \tag{A.59}$$

Strain can be expressed as $\delta = \sigma/Y$. Substituting the expression for stress from equation A.56 yields the following expression for strain.

$$\delta = m(\ddot{y} + \ddot{z})\frac{b}{4Y_c I}\left(4l_b + 3l_m\right) \tag{A.60}$$

Equation A.60 can be easily solved for the term $m(\ddot{y} + \ddot{z})$ and substituted into equation A.59. The resulting expression for vertical deflection is shown in equation A.61.

$$z = \frac{l_b}{3b}\left(l_c + l_b\right)\delta \tag{A.61}$$

$b^*$, which relates strain to vertical displacement, or $z = \delta/b^*$, can then be expressed as:

$$b^* = \frac{3b}{l_b(l_c + l_b)} \tag{A.62}$$

As previously, the effective spring constant, $k_{sp}$, is equal to $Yb^*b^{**}$. Substituting these new expressions for $b^*$ and $b^{**}$ into the models in equation A.30 and A.33 will yield a dynamic model incorporating a pin-pin mounting.

# Appendix B: Analytical Model of an Electrostatic Generator

Chapter 6 discussed the modeling and design of electrostatic converters. It was shown that the in-plane electrostatic converter was the topology best suited to the current application. A model was presented for all types of electrostatic converters, however the detailed derivation of the models was not presented in chapter 6 in order to improve readability. The purpose of this appendix is to present the detailed derivation of the model for the in-plane gap closing converter. The derivation of the other two types of converters follow the exact same procedure outlined here. Additionally, an algorithm for simulation in Matlab is presented.

## 1. DERIVATION OF ELECTRICAL AND GEOMETRIC EXPRESSIONS

A schematic of the in-plane gap closing converter is shown here as Figure B.1. The simple conversion circuit used, shown previously as Figure 6.1, is repeated here as Figure B.2. The basis of conversion is the variable capacitor, $C_v$ in Figure B.2. The capacitor is formed between the interdigitated combs shown in Figure B.1. The center shuttle moves left to right when excited by vibrations, changing the dielectric gap between the comb fingers, and thus changing the capacitance. For a more complete explanation of the process, see chapter 6.

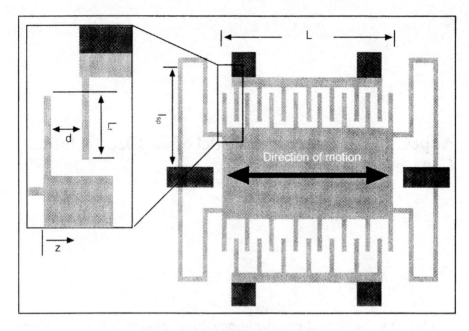

*Figure –B.1.* Schematic of in-plane gap closing converter.

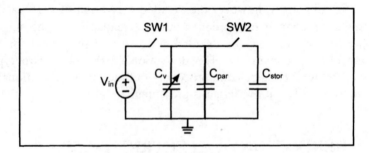

*Figure –B.2.* Simple conversion circuit for electrostatic converters.

Let us first develop the expression for energy converted per cycle from the circuit in Figure B.2. The capacitor $C_v$ is oscillating between a maximum value, $C_{max}$, and a minimum value $C_{min}$. When $C_v$ is at $C_{max}$, SW1 is closed so that the voltage, $V$, across $C_v$ is $V_{in}$. The charge, $Q$, on $C_v$ at this point is $Q = (C_{max} + C_{par})V_{in}$ where $C_{par}$ is a parasitic capacitance. As the center shuttle moves from the maximum capacitance position to the minimum capacitance position, both switches are open, so the charge on $C_v$ is constant. Therefore, at the minimum capacitance position $Q = (C_{min} + C_{par})V_{max}$ where $V_{max}$ is the voltage at that position. These expressions are summarized in equation B.1.

$$Q = \left(C_{max} + C_{par}\right)V_{in} = \left(C_{min} + C_{par}\right)V_{max} \qquad \text{(B.1)}$$

Solving for $V_{max}$ yields:

$$V_{max} = \frac{C_{max} + C_{par}}{C_{min} + C_{par}}V_{in} \qquad \text{(B.2)}$$

The energy stored in the variable and parasitic capacitors is greater at $C_{min}$ than at $C_{max}$ because mechanical work was done to move the $C_v$ from the maximum to minimum position increasing the total energy stored in the capacitor. So, the energy gain per cycle is given by equation B.3.

$$E = \frac{1}{2}\left(C_{min} + C_{par}\right)V_{max}^2 - \frac{1}{2}(C_{max} + C_{par})V_{in}^2 \qquad \text{(B.3)}$$

Substituting equation B.2 in to B.3 yields:

$$E = \frac{1}{2}\left(C_{min} + C_{par}\right)\left(\frac{C_{max} + C_{par}}{C_{min} + C_{par}}\right)^2 V_{in}^2 - \frac{1}{2}(C_{max} + C_{par})V_{in}^2 \qquad \text{(B.4)}$$

Equation B.4 can be algebraically manipulated to yield the following expression:

$$E = \frac{1}{2}V_{in}^2\left(C_{max} + C_{par}\right)\left(\frac{C_{max} + C_{par}}{C_{min} + C_{par}} - 1\right) \qquad \text{(B.5)}$$

Equation B.5 can be further reduced to yield the expression in equation B.6a, which is the same expression presented as equation 6.1a.

$$E = \frac{1}{2}V_{in}^2\left(C_{max} + C_{par}\right)\left(\frac{C_{max} + C_{min}}{C_{min} + C_{par}}\right) \qquad \text{(B.6a)}$$

If $V_{max}$ is an important parameter because of a particular switch implementation, equation B.6a can be rewritten as equation B.6b by substitution of the expression in equation B.2. Equation B.6b is the same equation presented previously as equation 6.1b.

$$E = \frac{1}{2}V_{max}V_{in}(C_{max} - C_{min}) \tag{B.6b}$$

The capacitance of a structure is given by $C = \varepsilon_0 A/d$ where $\varepsilon_0$ is the dielectric constant of free space, $A$ is the overlap area of the electrodes, and $d$ is the distance between the electrodes. The capacitance of the structure, $C_v$, shown in Figure B.1 is then given by equation B.7.

$$C_v = N_g \varepsilon_0 L_f h \left( \frac{1}{d-z} + \frac{1}{d+z} \right) \tag{B.7}$$

where, $N_g$ is the number of dielectric gaps per side of the device, $\varepsilon_0$ is the dielectric constant of free space, $L_f$ is the overlapping length of the fingers as shown in Figure B.1, $h$ is the thickness of the device, $d$ is the nominal gap between fingers as shown in Figure B.1, $z$ is the deflection of the springs as shown in Figure B.1.

Equation B.7 can be algebraically reduced to the equation for capacitance presented in chapter 6 as equation 6.14 and shown here as equation B.8.

$$C_v = N_g \varepsilon_0 L_f h \left( \frac{2d}{d^2 - z^2} \right) \tag{B.8}$$

The minimum capacitance occurs at the nominal position where $z$ equals zero. Setting $z$ equal to zero in equation B.8 yields:

$$C_{min} = \frac{2N_g \varepsilon_0 L_f h}{d} \tag{B.9}$$

If it is assumed that mechanical stops are implemented to limit the structure to a maximum displacement of $z_{max}$, where $z_{max}$ is less than $d$, the minimum dielectric gap is $d - z_{max}$. The maximum capacitance can then be written as:

$$C_{max} = N_g \varepsilon_0 L_f h \left( \frac{2d}{d^2 - z_{max}^2} \right) \tag{B.10}$$

## 2. DERIVATION OF MECHANICAL DYNAMICS AND ELECTROSTATIC FORCES

A quick estimate of power generated can be derived from equations B.6a, B.9, and B.10. Note that power for this device is given by $P = 2fE$ where $f$ is the driving frequency of the vibrations. The factor of 2 comes from the fact that the structure undergoes two electrical cycles for each mechanical cycle. Therefore power can be estimated purely from geometric design parameters, physical constants, and the input voltage. This method of estimating power was used to generate the data shown in chapter 6 as Figure 6.4. However, there is a large assumption underlying this analysis. It is assumed that the driving vibrations do actually drive the center shuttle of the structure hard enough to physically reach the mechanical stops under the specified conditions (i.e. input voltage, nominal dielectric gap, etc.). A full and accurate analysis needs to include the mechanical dynamics of the system so that if the fluid damping and electrostatic forces are such that the shuttle mass does not reach the mechanical stops this will come out in the analysis and simulation.

Consider the general coupled dynamic equation of a general oscillating mass system given in Chapter 6 as equation 6.2 and repeated here as equation B.11.

$$m\ddot{z} + f_e() + f_m() + kz = -m\ddot{y} \tag{B.11}$$

where $m$ is the mass of the oscillating capacitive structure, $k$ is the stiffness of the flexures on the capacitive structure, $z$ is the displacement of the capacitive structure, $y$ is the input vibration signal, $f_e(\ )$ represents the electrically induced damping force function, and $f_m(\ )$ represents the mechanical damping force function.

The mass, $m$, is simply the mass of the center shuttle and any additional mass attached to the center mass. There is a folded flexure at each corner of the center shuttle. Each element of the folded flexure can be treated as a fixed – fixed beam. The effective spring constant for a fixed – fixed beam is given by equation B.12.

$$k_{fixed-fixed} = \frac{12YI}{l_{sp}^3} \tag{B.12}$$

where $Y$ is the elastic constant, or Young's modulus, $I$ is the moment of inertia of the beam, and $l_{sp}$ is the length of the fixed-fixed beam, see Figure B.1.

The folded flexure element can be treated as two springs in series each with an effective spring constant of the fixed-fixed beam. Thus the stiffness of one flexure is half that of the fixed-fixed beam. Note that the two beams that make up the flexure appear to be quite different lengths in Figure B.1, however, the figure is not to scale and in reality the two beams are almost exactly the same length. Four folded flexures act in parallel on the center shuttle. Therefore, the aggregate spring constant, the $k$ term in equation B.11, is four times the spring constant of a single folded flexure or 2 times the spring constant of a fixed-fixed beam. The expression for the aggregate spring constant then becomes:

$$k = \frac{24YI}{l_{sp}^3} \qquad (B.13)$$

Derivation of the expressions for the electrostatic force, $f_e()$, and mechanical damping force, $f_m()$, require a little more work. First consider the mechanical damping force, $f_m()$. The dominant damping mechanism is fluid damping. The shear force of the fluid, air in this case, between two flat surfaces moving in parallel is usually referred to as Couette damping. In this case, Couette damping acts between the large center shuttle and the substrate beneath it. The expression for the force exerted on the shuttle mass is given by equation B.14.

$$F_{coutte} = \frac{\mu A}{d_0} \dot{z} \qquad (B.14)$$

where $\mu$ is the viscosity of air (18 microPascal seconds), $A$ is the area of the of the center shuttle, and $d_0$ is the distance between the center shuttle and the substrate beneath it.

A second fluid damping mechanism is also active. As comb fingers move closer together, the air between them is compressed and exerts a force opposing the motion. This damping force is usually referred to as squeeze-film damping. The general expression for the force exerted on a rectangular plate moving toward or away from another plate is given by:

$$F_{fluid} = \frac{16\mu l w^3}{x^3} \dot{x} \qquad (B.15)$$

where $l$ is the length of the rectangular plate, $w$ is the width of the rectangular plate, and $x$ is the distance between the plates.

Applying equation B.15 to the in-plane gap closing converter yields the following expression for the squeeze film damping force.

$$F_{squeeze-film} = 16N_g \mu L_f h^3 \left( \frac{1}{(d-z)^3} + \frac{1}{(d+z)^3} \right) \dot{z}$$ (B.16)

The total damping force is just $f_m() = F_{coette} + F_{squeeze-film}$, given in equation B.17, which is the same expression presented in chapter 6 as equation 6.13.

$$f_m() = \left( \frac{\mu A}{d_0} + 16N_g \mu L_f h^3 \left( \frac{1}{(d-z)^3} + \frac{1}{(d+z)^3} \right) \right) \dot{z}$$ (B.17)

Next consider the electrostatic force $f_e()$. The electrostatic force on a body as a function of position is given by:

$$F = -\left( \frac{\partial U}{\partial x} \right)$$ (B.18)

where $U$ is the electrostatic energy stored, and $x$ is a general displacement variable.

The energy stored in a capacitive device is given by:

$$U = \frac{1}{2}CV^2 = \frac{Q^2}{2C}$$ (B.19)

Either form of equation B.19 can be used with equation B.18. However, it is generally better to use the expression that will simplify the mathematics as much as possible. In this case, the charge, $Q$, is held constant and so is not a function of displacement whereas $V$ is a function of displacement. Therefore, the second form of equation B.19 will be used. The energy term for the in-plane gap closing converter, using the capacitance expression in equation B.8, is as follows:

$$U = \frac{Q^2(d^2 - z^2)}{4N_g \varepsilon_0 L_f hd}$$ (B.20)

Applying equation B.20 to equation B.18 using $z$ as the displacement variable yields:

$$f_e() = \frac{Q^2 z}{2N_g d\varepsilon_0 L_f h} \tag{B.21}$$

Equation B.21 is the same expression presented as equation 6.15 in chapter 6.

## 3.    SIMULATION OF THE IN-PLANE GAP CLOSING CONVERTER

All of the equations necessary to simulate the system have now been derived. The model of the system is comprised of equations B.6a, B.9, B.10, B.11, B.13, B.17 and B.21. Making substitutions, these seven equations can be written as the following two coupled equations.

$$m\ddot{z} + \frac{Q^2}{2N_g d\varepsilon_0 L_f h} z + \left( \frac{\mu A}{d_0} + 16 N_g \mu L_f h^3 \left( \frac{1}{(d-z)^3} + \frac{1}{(d+z)^3} \right) \right) \dot{z}$$

$$+ \frac{24YI}{l_{sp}^3} z = -m\ddot{y} \tag{B.22}$$

$$E = V_{in}^2 \left( N_g \varepsilon_0 L_f h \left( \frac{2d}{d^2 - z_{max}^2} \right) + C_{par} \right) \left( \frac{2d^2 - z_{max}^2}{d^2 - z_{max}^2} \right) \left( \frac{N_g \varepsilon_0 L_f h}{C_{par} d + N_g \varepsilon_0 L_f h} \right) \tag{B.23}$$

Simulation of this system is not quite as straight forward as simulation of the piezoelectric converter. There are two effects that contribute to this difficulty.    The first is that at certain times the switches close, instantaneously (at least compared to the mechanical system) changing some of the values that are considered to be constant for purposes of the mechanical simulation, such as the charge $Q$. The second is the impact of the center shuttle mass with the mechanical stops. This impact is modeled as a purely elastic impact with a coefficient of restitution of 0.5 (Lee and

Pisano 1993). The Matlab differential equation solvers cannot model and solve these two effects. The following is the outline of the procedure that was used to simulate the system in Matlab.

First, the geometric dimensions, physical constants, and initial conditions are specified. A differential equation solver is then used to solve equation B.22. The Matlab function 'ode23s' seems to work the best. The dynamic simulation of equation B.22 needs to be stopped so that switch 1 can be closed if $C_v = C_{max}$ and so that switch 2 can be closed if $C_v = C_{min}$. Note that ideal switches are assumed for simulation purposes. The Matlab differential equation solvers can monitor the state variables and simple functions of the state variables for zero crossings (that is where a state variable or function of that variable crosses zero), and terminate the solver if at zero crossings. The case where $C_v = C_{max}$ occurs when $\dot{z}$ is zero or when the center mass hits the mechanical stops ($|z| - z_{max\_allowed} = 0$). The case where $C_v = C_{min}$ occurs when the displacement z equals zero (i.e. the center mass is in the center position). When any of these three conditions occur, the differential equation solver stops, circuit calculations are made and stored, and the solver is restarted using the current state values and time as the initial conditions. The following list outlines specifically what is done when each of the three conditions occurs.

- Condition: $\dot{z} = 0$. $C_v$ is at $C_{max}$. Close switch 1.
  - For each time step returned by the solver (that is all time values since the last solver last stopped), calculate electrical circuit values.
    - Calculate $C_v$ using equation B.7.
    - Both switches have been open, so $Q$ is constant.
    - Calculate voltage on $C_v$ as $V = Q / (C_v + C_{par})$.
  - Close switch 1 and calculate new (current) circuit values.
    - $V = V_{in}$.
    - More charge is put on variable capacitor. Re-calculate current charge as $Q = (C_v + C_{par})V_{in}$. Note the current value of $C_v$ is $C_{max}$.
    - Calculate the amount of charge put in, $Q_{in}$, as $Q$ minus the charge that was left on the capacitor just before the charge was re-calculated.
  - Append state variables returned by the solver and circuit values to the end of a persistent storage matrix.
  - Set initial conditions to the current state values and restart differential equation solver.
- Condition: $|z| - z_{max\_allowed} = 0$. Center shuttle hit the mechanical stops. $C_v$ is at $C_{max}$. Close switch 1.

- o Procedures are exactly the same as those listed above except that a new value for velocity, $\dot{z}_{new} = -0.5\ \dot{z}_{old}$, needs to be set as the initial condition before the solver restarts. In this way the effect of the impact is included.
- Condition: $z = 0$. $C_v$ is at $C_{min}$. Close switch 2.
  - o For each time step returned by the solver, calculate electrical circuit values as detailed above.
  - o Close switch 2 and calculate new (current) circuit values.
    - When switch 2 closes the voltage across the variable capacitor is shorted with the output voltage. Calculate new voltages as $V = V_{stor} = (Q + Q_{stor}) / (C_v + C_{par} + C_{stor})$.
    - Calculate new value for charge on output capacitor as $Q_{stor} = C_{stor}V_{stor}$.
    - Calculate amount of charge left on variable capacitor as $Q = (C_v + C_{par})V$.
  - o Append state variables returned by the solver and circuit values to the end of a persistent storage matrix.
  - o Set initial conditions to the current state values and restart differential equation solver.

After the sequence outlined above has been run for a predetermined amount of time or number of iterations, the energy gained versus time can then be calculated as the increased energy on the storage capacitor minus the amount of energy put onto the variable capacitor. This simulation can, and has, been used as the objective function for an optimization algorithm to determine design parameters such as length of fingers, nominal gap, size of center shuttle, etc. After a number of simulation iterations it will become clear that the best designs are ones in which the center shuttle mass just barely reaches the mechanical stops. If the electrostatic forces are too high, and the center shuttle does not come very close to reaching the stops, the maximum capacitance will not be as high as it could, and the resulting energy gain per cycle is not very high. If the electrostatic forces are too low the center mass will ram into mechanical stops. This is not ideal because energy is dissipated in the impact, and because higher electrostatic forces would result in more work going into the conversion of mechanical kinetic energy to electrostatic energy. Thus, the best designs are ones in which the target vibrations drive the center shuttle just hard enough to barely reach the mechanical stops.

# References

Amirtharajah R., Chandrakasan A.P., 1998. Self-Powered Signal Processing Using Vibration-Based Power Generation. *IEEE Journal of Solid State Circuits*, Vol. 33, No. 5, p. 687-695.

Amirtharajah R., 1999. *Design of Low Power VLSI Systems Powered by Ambient Mechanical Vibration*. Ph.D Thesis, Department of Electrical Engineering, Massachusetts Institute of Technology, June 1999.

Amirtharajah R., Meninger S., Mur-Miranda J. O., Chandrakasan A. P., Lang J., 2000. A Micropower Prgrammable DSP Powered using a MEMS-based Vibration-to-Electric Energy Converter. *IEEE Interational Solid State Circuits Conference*, 2000. p. 362-363.

Bates J.B., Dudney N.J., Neudecker B., Ueda A., Evans C.D., 2000. Thin-film lithium and lithium-ion batteries. *Solid State Ionics* 135: 33-45.

Bellew C., 2002. *An SOI Process for Integrated Solar Power, Circuitry and Actuators for Autonomous Microelectromechanical Systems*, Ph.D Thesis, Department of Electrical Engineering, University of California at Berkeley, May 2002.

Beer F.P., Johnston E.R., 1992. *Mechanics of Materials*, McGraw-Hill, Inc., 1992.

Chandrakasan A., Amirtharajah R., Goodman R. J., Rabiner W., 1998. Trends in low power digital signal processing. *Proceedings of the 1998 IEEE International Symposium on Circuits and Systems*, 1998, p. 604-7.

Chang Y.T., Lin L., 2001. Localized Silicon Fusion and Eutectic Bonding for MEMS Fabrication and Packaging. *Journal of Microelectromechanical Systems*, vol. 9 (no. 1) (2001) 3 – 8.

Cragun, R., Howell, L.L., 1999. Linear thermomechancial microactuators. *Proc. ASME IMECE MEMS* (1999) pp. 181-188.

Cronos 2003. www.memsrus.com, 2003.

Crossbow, 2003. www.xbow.com/Products/Wireless_Sensor_Networks.htm, 2003.

Davis W. R., Zhang N., Camera K., Chen F., Markovic D., Chan N., Nikolic B., Brodersen R. W., 2001. A design environment for high throughput, low power dedicated signal processing systems. *Proceedings of the IEEE 2001 Custom Integrated Circuits Conference*, 2001, p. 545-8.

Doherty L., Warneke B.A., Boser B.E., Pister K.S.J., 2001. Energy and performance considerations for smart dust. *International Journal of Distributed Systems & Networks*, vol. 4, no.3, 2001, pp. 121-33.

Dust, 2003. www.dust-inc.com, 2003.

Economist, 1999. Power from the people. *The Economist,* Apr 15, 1999.

Ember, 2003. www.ember.com, 2003.

Epstein AH, et al., 1997. Micro-Heat Engine, Gas Turbine, and Rocket Engines – The MIT Microengine Project. AIAA 97-1773, *28th AIAA Fluid Dynamics Conf.,* Snowmass Village, CO, June 1997

Evans J. G., Shober R. A., Wilkus S. A., Wright G. A., 1996. A low-cost radio for an electronic price label system. *Bell Labs Technical Journal* vol. 1, Issue 2, 1996. pp. 203 – 215.

Flynn A.M., Sanders S.R. 2002. Fundamental limits on energy transfer and circuit considerations for piezoelectric transformers. *IEEE Transactions on Power Electronics,* vol.17, (no.1), IEEE, Jan. 2002. p.8-14.

Freeplay, 2003. www.freeplay.net, 2003.

Friedman D., Heinrich H., Duan D-W., 1997. A Low-Power CMOS Integrated Circuit for Field-Powered Radio Frequency Identification. *Proceedings of the 1997 IEEE Solid-State Circuits Conference,* p. 294 – 295, 474.

Fu K., Knobloch A.J., Martinez F.C., Walther D.C., Fernandez-Pello C., Pisano A.P., Liepmann D., 2001. Design and Fabrication of a Silicon-Based MEMS Rotary Engine, *ASME IMECE,* New York, November 11-16, 2001.

Gates B., 2002. The disappearing computer. *The Economist, Special Issue: The World in 2003,* December 2002, p. 99.

Glynne-Jones P., Beeby S. P., James E. P., White N. M., 2001. The modelling of a piezoelectric vibration powered generator for microsystems. *Transducers 01 / Eurosensors XV,* June 10 – 14, 2001.

Haartsen J.C., Mattison S., 2000. Bluetooth – A New Low-Power Radio Interface Providing Short-Range Connectivity. *Proceedings of the IEEE,* 88(10):1651 – 1661

Harb J., LaFollete R.M., Selfridge R.H., Howell L.L. 2002. Microbatteries for self-sustained hybrid micropower supplies. *Journal of Power Sources* 104: 46-51.

Hart R.W., White H.S., Dunn B., Rolison D.R., 2003. 3-D Microbatteries. *Electrochemistry Communications* 5:120-123

Heinzel A., Hebling C., Muller M., Zedda M., Muller C., 2002. Fuel cells for low power applications. *Journal of Power Sources* 105: 250 – 255

Hill J., Culler D., 2002. Mica: A Wireless Platform for Deeply Embedded Networks. *IEEE Micro* 22(6):12-24.

Hitachi, 2003. Hitachi Unveils Smallest RFID Chip. *RFiD Journal,* March 14, 2003.

Holloday J.D., Jones E.O., Phelps M., Hu J., 2002. Microfuel processor for use in a miniature power supply. *Journal of Power Sources* 108:21-27

Hsu, T-R, 2000. Packaging Design of Microsystems and Meso-Scale Devices. *IEEE Transactions on Advanced Packaging,* vol. 23 (no. 4) (2000) 596 – 601.

Ikeda, T., 1990. *Fundamentals of piezoelectricity,* Oxford University Press, New York, 1990.

Iowa Thin Film Technologies, 2003. www.iowathinfilm.com , Jan. 2003.

Isomura K., Murayama M., Yamaguchi H., Ijichi N., Asakura H., Saji N., Shiga O., Takahashi K., Tanaka S., Genda T., Esashi M., 2002. Development of Microturbocharger and Microcombustor for a Three-Dimensional Gas Turbine at Microscale. *ASME IGTI 2002 TURBO EXPO,* Paper GT-2002-30580, Amsterdam, Netherlands, June 6, 2002

Jaeger, R. C., 1993. *Introduction to Microelectronic Fabrication,* Addison Wesley Publishing Company, Inc., 1993.

James, M.L., Smith, G.M., Wolford, J.C., Whaley, P.W., 1994. *Vibration of Mechanical and Structural Systems,* Harper Collins College Publishers, New York, NY 1994.

Kang, S., Lee, S-J. J., Prinz, F.B., 2001. Size does matter: the pros and cons of miniaturization. *ABB Review*, 2001, no.2, p.54-62.

Kassakian, J.C., Schlecht, M.F., Verghese, G.C., 1992. *Principles of Power Electronics.* Addison Wesley Publishing Company, 1992.

Kordesh K., Simader G., 2001. Fuel cells and their applications. VCH Publishers, New York

Laerme F, Schilp A, Funk K, Offenberg M., 1999. Bosch deep silicon etching: improving uniformity and etch rate for advanced MEMS applications. *Twelfth IEEE International Conference on Micro Electro Mechanical Systems* 1999, pp.211-16.

Lee S. H., 2001. Development of high-efficiency silicon solar cells for commercialization. *Journal of the Korean Physical Society*, vol.39, no.2, Aug. 2001, pp.369-73.

Lee C., Arslan S., Liu Y-C., Fréchette L.G., 2003. Design of a Microfabricated Rankine Cycle Steam Turbine for Power Generation. *ASME IMECE*, Washington, D.C., Nov. 16-21, 2003

Lee A. P., and Pisano A. P., 1993. Repetitive impact testing of micromechanical structures. *Sensors and Actuators A (Physical)*, A39 (1) (1993) 73-82.

Lee S.S., and White R.M., 1995. Self-excited Piezoelectric Cantilever Oscillators. *Proc. Transducers 95/Eurosensors IX*, (1995) 41 – 45.

Lesieutre G.A., 1998. Vibration damping and control using shunted piezoelectric materials. *Shock and Vibration Digest* vol. 30, pp 187-195.

Li H., Lal M., 2002. Self-reciprocating radio-isotope powered cantilever, *Journal of Applied Physics* 92(2):1122 – 1127.

Lott C. D., 2001, *Electrothermomechanical Modeling of a Surface-Micromachined Linear Displacement Microactuator*, M.S. Thesis, Department of Mechanical Engineering, Brigham Young University, August 2001.

Lott C.D., McLain T.W., Harb J.N., Howell L.L., 2002. Modeling the thermal behavior of a surface-micromachined linear-displacement thermomechanical microactuator. *Sensors & Actuators A-Physical*, vol.A101, no.1-2, 30 Sept. 2002, pp.239-50.

Madou M., 1997. *Fundamentals of Microfabrication*, CRC Press LLC, Boca Raton Florida, 1997.

Maluf N., 2000. *An Introduction to Microelectromechanical Systems Engineering*, Artech House, Inc., Norwood, Massachusetts, 2000.

Maxwell , 2003. www.maxwell.com/ultracapacitors/, 2003.

Mehra A., Zhang X., Ayon A. A., Waitz I. A., Schmidt M. A., Spadaccini C. M., 2000. A Six-Wafer Combustion System for a Silicon Micro Gas Turbine Engine. *Journal of Microelectromechanical Systems*, 9 (4) (2000) 517-526.

Meninger S., Mur-Miranda J. O., Amirtharajah R., Chandrakasan A. P., Lang, J., 1999, Vibration-to-Electric Conversion, *ISPLED99* San Diego, CA, USA, p. 48 – 53.

Meninger S., Mur-Miranda J.O., Amirtharajah R., Chandrakasan A.P., Lang J.H., 2001. Vibration-to-Electric Energy Conversion. *IEEE Trans. VLSI Syst.*, 9 (2001) 64-76.

Moulson A.J., Herbert J.M., 1997. *Electroceramics Materials Properties Applications.* Chapman and Hall, 1997.

National Research Council, Rowe J.E. 1997. *Energy-Efficient Technologies for the Dismounted Soldier*, National Academy Press, 1997.

NEC, 2003. www.nec-tokin.net/now/english/product/ hypercapacitor/outline02.html, 2003.

Nijs J.F., Szlufcik J., Poortmans J., Mertens R.P., 2001. Crystalline silicon based photovoltaics: technology and market trends. *Modern Physics Letters B*, vol.15, no.17-19, 20 Aug. 2001, pp.571-8.

Otis B., Rabaey J., 2002. A 300µW 1.9GHz CMOS Oscillator Utilizing Micromachined Resonators., *Proceedings of the 28th European Solid State Circuits Conference*, Florence Italy, September 24 – 26, 2002.

Ottman G. K., Hofmann H. F., Lesieutre G. A. 2003. Optimized piezoelectric energy harvesting circuit using step-down converter in discontinuous conduction mode. *IEEE Transactions on Power Electronics*, vol.18, no.2, 2003, pp.696-703.

Ottman G. K., Hofmann H. F., Bhatt A. C., Lesieutre G. A. 2002. Adaptive piezoelectric energy harvesting circuit for wireless remote power supply. *IEEE Transactions on Power Electronics*, vol.17, no.5, 2002, pp.669-76.

Park S.E., and Shrout T.R. 1997. Characteristics of Relaxor-Based Piezoelectric Single Crystals for Ultrasonic Transducers. *IEEE Trans. on Ultrasonics, Ferroelectric and Frequency Control Special Issue on Ultrasonic Transducers*, Vol. 44, No. 5, 1140-1147 (1997).

Pescovitz D., 2002. The Power of Small Tech. Smalltimes, Vol. 2, No. 1, 2002.

Piezo Systems, Inc., 1998. *Piezoelectric Actuator/Sensor Kit Manual*, Piezo Systems, Inc., Cambridge MA., (1998).

Rabaey J., Ammer J., Karalar T., Li S., Otis B., Sheets M., Tuan T., 2002. PicoRadios for Wireless Sensor Networks: The Next Challenge in Ultra-Low-Power Design. *Proceedings of the International Solid-State Circuits Conference*, San Francisco, CA, February 3-7, 2002.

Rabaey J. M., Ammer M. J., da Silva J. L., Patel D., Roundy S., 2000. PicoRadio Supports Ad Hoc Ultra-Low Power Wireless Networking. *IEEE Computer*, Vol. 33, No. 7, p. 42-48.

Raible C., Michel H., 1998. Bursting with power. *Siemens Components* (English Edition), vol.33, (no.6), Siemens AG, Dec. 1998. p.28-9

Randall J. F. *On ambient energy sources for powering indoor electronic devices*, Ph.D Thesis, Ecole Polytechnique Federale de Lausanne, Switzerland, May 2003.

Riehl P. S., Scott K., L., Muller R. S., Howe R., T., 2002. High-Resolution Electrometer with Micromechanical Variable Capacitor. *Solid State Sensor, Actuator and Microsystems Workshop*, Hilton Head Island, South Carolina, June 2002, pp. 305 – 308.

Rosenber R. C., Karnopp D. C. 1983. *Introduction to Physical System Dynamics*, McGraw Hill Inc., 1983.

Roundy S., Wright P. K., Pister K. S. J., 2002. Micro-Electrostatic Vibration-to-Electricity Converters, *ASME IMECE*, Nov. 17-22, 2002, New Orleans, Louisiana.

Roundy S., Otis B., Chee, Y-H., Rabaey J., Wright P.K., 2003. A 1.9 GHz Transmit Beacon using Environmentally Scavenged Energy. *ISPLED 2003*, August 25-27, 2003, Seoul Korea.

Santavicca D., Sharp K., Hemmer J., Mayrides B., Taylor D., Weiss J., 2003. A Solid Piston Micro-engine for Portable Power Generation. *ASME IMECE*, Washington, D.C., Nov. 16-21, 2003.

Schittowski K., 1985. NLQPL: A FORTRAN-Subroutine Solving Constrained Nonlinear Programming Problems. *Annals of Operations Research*, Vol. 5, pp. 485-500, 1985.

Schmidt V.H., Theoretical Electrical Power Output per Unit Volume of PVF2 and Mechanical-to-Electrical Conversion Efficiency as Functions of Frequency. *Proceedings of the Sixth IEEE International Symposium on Applications of Ferroelectrics*, (1986) 538-542.

Seiko, 2003. www.seikowatches.com, 2003.

Shearwood C., Yates R.B., 1997. Development of an electromagnetic micro-generator, *Electronics Letters*, vol.33, (no.22), IEE, 23 Oct. 1997. p.1883-4

Shenck N. S., Paradiso J. A., 2001. Energy Scavenging with Shoe-Mounted Piezoelectrics, *IEEE Micro*, 21 (2001) 30-41.

Sim W. Y., Kim G. Y., Yang S. S., 2001. Fabrication of micro power source (MPS) using a micro direct methanol fuel cell (μDMFC) for the medical application. *Technical Digest. MEMS 2001* p. 341-344.

Smith A.A., 1998. Radio frequency principles and applications : the generation, propagation, and reception of signals and noise, IEEE Press, New York

Srinivasan U., Liepmann D., Howe R.T., 2001. Microstructure to substrate self-assembly using capillary forces, *Journal of Microelectromechanical Systems,* 10-1, 17-24 (2001).

Starner T., 1996. Human-powered wearable computing. *IBM Systems Journal,* 35 (3) (1996) 618-629.

Stordeur M., Stark I., 1997. Low Power Thermoelectric Generator – self-sufficient energy supply for micro systems. *16th International Conference on Thermoelectrics,* 1997, p. 575 – 577.

Tang W.C., Nguyen T.-C.H., Howe R.T., 1989. Laterally driven polysilicon resonant microstructures. *Sensors Actuators,* vol. 20, pp. 25–32, 1989.

Toshiba, 2003. www.toshiba.co.jp/about/press/2003_03/pr0501.htm, 2003.

TRS Ceramics, 2003. www.trsceramics.com, Jan. 2003.

Tzou H. S., *Piezoelectric Shells, Distributed Sensing and Control of Continua,* Kluwer Academic Publishers, Norwell, Massachusetts, 1993.

Verardi P, Craciun F, Dinescu M. 1997. Characterization of PZT thin film transducers obtained by pulsed laser deposition. *IEEE Ultrasonics Symposium Proceedings.* vol.1, 1997, pp.569-72.

Wang D., Arens E., Webster T., Shi M., 2002. How the Number and Placement of Sensors Controlling Room Air Distribution Systems Affect Energy Use and Comfort, *International Conference for Enhanced Building Operations,* Richardson, TX, October 2002.

Warneke B. Atwood B. Pister K.S.J., 2001. Smart Dust Mote Forerunners, *Fourteenth Annual International Conference on Microelectromechanical Systems (MEMS 2001),* Interlaken, Switzerland, Jan. 21-25, 2001.

Whalen S., Thompson M., Bahr D., Richards C., Richards R., 2003. Design, Fabrication and Testing of the P3 Micro Heat Engine. Sensors and Actuators 104(3):200-208.

Williams C.B, Yates R.B., 1995. Analysis of a micro-electric generator for Microsystems. *Transducers 95/Eurosensors IX,* (1995) 369 – 372.

Williams C.B., Shearwood C., Harradine M.A., Mellor P.H., Birch T.S., Yates R.B., 2001. Development of an electromagnetic micro-generator. *IEEE Proceedings-Circuits, Devices and Systems,* vol.148, (no.6), IEE, Dec. 2001. p.337-42

Xsilogy, 2003. www.xsilogy.com, 2003.

# Index